The Elements of Architecture

The Elements of Architecture

Principles of Environmental Performance in Buildings

Scott Drake

Illustrations by Adam Brown and Tristan Wong

publishing for a sustainable future

London • Sterling, VA

First published by Earthscan in the UK and USA in 2009

This is an internationalized version of a book first published by University of New South Wales Press in 2007 under the title *The Third Skin: Architecture, Technology and Environment*

ISBN: 978-1-84407-716-8 hardback
          978-1-84407-717-5 paperback

Typeset by Safehouse Creative
Cover design by Yvonne Booth

For a full list of publications please contact:

Earthscan
Dunstan House
14a St Cross St
London, EC1N 8XA, UK
Tel: +44 (0)20 7841 1930
Fax: +44 (0)20 7242 1474
Email: earthinfo@earthscan.co.uk
Web: **www.earthscan.co.uk**

22883 Quicksilver Drive, Sterling, VA 20166-2012, USA

Earthscan publishes in association with the International Institute
for Environment and Development

A catalogue record for this book is available from the British Library

Library of Congress Cataloging-in-Publication Data
Drake, Scott, Dr.
  The elements of architecture : principles of environmental performance in buildings / Scott Drake.
     p. cm.
  Includes bibliographical references and index.
  ISBN 978-1-84407-716-8 (hardback) – ISBN 978-1-84407-717-5 (pbk.) 1. Architecture–Environmental aspects. 2. Environmental engineering. I. Title.
  NA2542.35.D73 2009
  720.47–dc22

                                                                    2008040379

At Earthscan we strive to minimize our environmental impacts and carbon footprint through reducing waste, recycling and offsetting our CO_2 emissions, including those created through publication of this book. For more details of our environmental policy, see www.earthscan.co.uk.

This book was printed in the UK by TJ International, an ISO 14001 accredited company. The paper used is FSC certified and the inks are vegetable based.

Mixed Sources
Product group from well-managed forests and other controlled sources
www.fsc.org Cert no. SGS-COC-2482
© 1996 Forest Stewardship Council

FSC

contents

list of figures and tables

Figures

Tables

In normal contexts, the room, the simplest form of shelter, expresses the most benign potential of human life. It is, on the one hand, an enlargement of the body: it keeps warm and safe the individual it houses in the same way the body encloses and protects the individual within; like the body, its walls put boundaries around the self preventing undifferentiated contact with the world, yet in its windows and doors, crude versions of the senses, it enables the self to move out into the world and allows that world to enter. But while the room is a magnification of the body, it is simultaneously a miniaturization of the world, of civilization. Although its walls, for example, mimic the body's attempt to secure for the individual a stable internal space – stabilizing the temperature so the body spends less time in this act; stabilizing the nearness of others so that the body can suspend its rigid and watchful postures; acting in these and other ways like the body so that the body can act less like a wall – the walls are also, throughout all this, independent objects, objects which stand apart from and free of the body, objects which realize the human being's impulse to project himself out into a space beyond the boundaries of the body in acts of making, either physical or verbal, that once multiplied, collected, and shared are called civilization.

Elaine Scarry, *The Body in Pain: The making and unmaking of the world*
Oxford University Press, New York and Oxford, 1985, p. 38

preface

Here is an image you may be familiar with: a young, twenty-something male, dressed in jeans and a T-shirt, sitting in the sun on a well-mown patch of grass, staring attentively at a laptop computer. The casual attire and access to a well-groomed landscape suggest that he is probably a university student, filling a bit of time between lectures by catching up on emails or organizing music files, or even, perhaps, studying lecture notes or preparing an assignment. This image, used by advertising agencies around the world, is intended to convince you that owning a laptop is not merely convenient, but is essential for a modern lifestyle where responsibility and leisure merge seamlessly together. The laptop computer is the ultimate product of digital technology – powerful, portable and versatile – that makes this combination of work and leisure possible.

But one thing intrigues me about those ads: where are the buildings? Is the laptop so convenient and user-friendly that you no longer need a desk and a chair, a power outlet, doors, windows, walls, floor or roof to provide the protection from the elements that is necessary for the work that the rest of us do every day? This idyllic garden – wireless, powerless and building-less – is something of an urban fantasy. For most of us, even on a university campus, gardens are for looking at or walking through on the way from one building to another. Although we might like to be in touch with nature on a regular basis, the fact is we spend nearly all of our time inside buildings. And while there is increasing concern for preserving the natural environment, the reality is that buildings, for the most part, *are* our environment.

While promoting advanced technology in the form of a laptop computer, these ads seem to imply that the most advanced form of building is no building at all. In this advertising fantasy, the building has disappeared altogether, leaving us in perfect harmony with the natural environment. The portable computer, along with MP3 players, digital cameras and mobile telephones, is the epitome of technology. These advanced tools enable us to do things that we cannot do with our body alone – talk to friends in different places, listen to music without someone in front of us to play it, or perform intricate manipulations of words, numbers and images. Buildings are sometimes thought of as 'technology' too, but since new buildings are often seen to be at odds with the more revered, historical styles that give our cities character and identity, rarely are they embraced with the same enthusiasm that is shown for other consumer goods.

Apart from a few high-tech exceptions, buildings are not normally thought of as a form of technology. This may be because buildings last a long time, together making up the living museum known as a city, where relics from past eras are still in use alongside recent works, setting standards that are used to judge and influence new designs. It may be because buildings need to accommodate human populations in ever-increasing numbers, and are thus resistant to the sort of miniaturization that goes on with gadgets such as computers and mobile telephones. The elements of walls, doors, windows and roofs relate at their most basic level to the size and shape of the human body, which has remained largely unchanged for many thousands of years. Or it may be that, even though there have been advances in construction and servicing of buildings, much of the technology used today is at least 100 years old, and probably more. Steel-frame construction, motorized elevators and glass curtain walls all date from the late 19th century, while materials such as plaster, bricks and tiles, still used in buildings today, date back at least to Roman times. Even the most high-tech building will still contain a a lot of low-tech material just to do the basic job of forming a strong and stable structure.

A building is perhaps the most basic form of technology, a made artefact designed to order, organize or change the shape of the world to make it more habitable for human beings. Buildings modify or adapt the natural environment, using a combination of low-tech and high-tech strategies to determine the physical qualities of interior space. In doing so, they alter the sensory interaction between the body and the natural environment. While the body is largely self-regulating, able to adjust to the ever-changing environmental conditions around it, humans have sought to extend the range of environments that can be inhabited, initially by adding an outer layer of clothing, and then by constructing another layer in the form of a building. This book is, in large part, about how the environment within buildings is created, and how that can be done in ways that provide greater connection with and respect for the natural environment. The aim is to show that environmental considerations are among the most fundamental of all design decisions, affecting the size, shape and orientation of any building, the materials used for construction, the interior layout, and the location of windows and doors that link inside and outside space.

Introduction:
ARCHITECTURE, TECHNOLOGY AND ENVIRONMENT

> I was born by Caesarean section, but you can't really tell ...
> except that when I leave my house, I always go out the window.
>
> Steven Wright

Buildings and nature

With the rise to prominence of the green movement in recent decades, the term *environment* has come to be associated with the natural ecosystems of the Earth, especially as they are increasingly threatened by human technological activity. But for architects, the word has long been used in a more general sense, to describe the various spaces inhabited by humans, both built and natural, such as in the terms *built environment* and *environmental design*. Architects are often thought of as the designers of buildings, but for many architects, a building is merely a means to create an environment in which people can live and work, and to determine the level of interaction with the world outside. Buildings modify nature, so that the people within can still enjoy sun, light and views, but without exposure to extremes of rain, wind or cold. The very reason buildings are designed and constructed is that the natural environment is unsuitable for a large number of human activities, except when modified by the walls, windows and roof of a building.

why we need build ing

For architects, the environmental movement raises some important questions. Is it possible for a profession involved in creating the built environment to help reduce or reverse the damage done to the natural environment? If buildings are essentially a means of adapting nature to suit human needs, is it even possible to make buildings that are *environmentally friendly*? The answers to these questions depend on a detailed understanding of how buildings modify nature in order to make spaces suitable for human habitation. To help understand how buildings modify nature, this book explores the idea of a building as an environmental *filter*, selectively allowing various aspects of the external environment to pass through to interior space.

The idea of the building as an environmental filter was proposed 60 years ago by James Marston Fitch, in his book *American Building*.[1] 'The central function of architecture', wrote Fitch, is 'to lighten the very stress of life'. By providing shelter, buildings reduce the need to constantly seek protection from the natural environment and, instead, free up time and energy to undertake the various social and cultural activities that characterize human achievement. These activities, according to Fitch, are the 'essence of human experience'.[2]

Of course, buildings do more than simply modify the external environment. They can express a sense of identity for individual owners or designers, or collectively for a society or a nation; they can be seen as manufactured objects produced by a building industry, or as manifestations of the technology of construction; they can act as investment for owners or developers speculating on rising values; or they can even be targets for military or terrorist campaigns, with devastating results. While each of these aspects is important, all have emerged from the fundamental and ongoing need to modify the environment in order to facilitate social and cultural activities that are essential to human life.

Modified environments

What is it about human culture that requires protection from the natural environment? Descriptions of the origins of architecture are many and varied, usually derived from either mythological narrative or anthropological speculation about 'primitive' cultures.[3] Writings about the first house by architects from Marcus Vitruvius to Le Corbusier reflect the combination of available material and technological ability; trees felled from a forest clearing, for example, could be used to make fire or to make shelter, or both together.[4]

An alternative explanation is that buildings became necessary with the move away from nomadic, hunter-gatherer lifestyles to permanent, agriculture-based settlements in temperate climates. The fixed location of crops made permanent dwellings viable, but it was the need to protect seed banks and livestock through winter that made buildings a necessity.[5] The temperature range to which humans had originally adapted was that of the African savannah and, although the move to higher latitudes made for more productive crops, it required strategies for dealing with the colder weather.[6]

Fortunately, even in the coldest of climates, the amount of environmental modification needed is relatively small. The average temperature on the surface of the Earth is about 15°C, thanks to an atmosphere that helps to retain the heat arriving from the sun. Without the atmosphere, the average temperature would be much lower, around −18°C, making it not just hard to breathe, but too cold to support most forms of life.[7] Because the Earth is roughly spherical, radiation from the sun is more intense at the equator than at the poles. But by circulating air masses and warm water currents, the atmosphere and oceans act to distribute heat across the temperate zones, reducing the extremes that would otherwise occur. This is also assisted by the *declination*, or 'tilt', of the Earth, which causes the apparent path of the sun to move back and forth across the equator throughout the course of a year. The result is that most of the temperatures experienced across the Earth's surface occur within a relatively narrow band, near or below normal human body temperature of 37°C.

Human environments

The body operates at 37°C in order to facilitate the various chemical reactions that are necessary to sustain life; in particular, the conversion of energy consumed as food into a form that can be used by muscles to do physical work. This constant body temperature can only be maintained through regular metabolism of calorific energy. Metabolism is a form of low-temperature combustion that is about 20 per cent efficient, which means that about 80 per cent of the energy must be dissipated to the environment around the body. The rate of heat loss for a human body is usually around 80 joules per second, which varies depending on the level of activity.

For the dissipation of heat to occur, a temperature gradient must exist between the human body and its immediate environment. If the outside temperature is too close to that of the body, or in fact rises above it, heat loss must be increased. This is usually done through evaporative cooling, made possible by sweat glands releasing water onto the surface of the skin. If the outside temperature drops too far below that of the body, physical activity may be required to increase the metabolic rate and thus the rate of heat production. But for a range of temperatures, between those that lead to sweating and to shivering, the rate of heat loss is adjusted by redirecting the flow of blood toward or away from the skin, without any effort being required. Known as *conductance adjustment*, this ability of the body to imperceptibly adjust the rate of heat loss means that we can be comfortable across a range of temperatures near or below that of the human body.

Of course, the temperature range that can be tolerated can be extended with a little ingenuity. Fire can provide a source of radiant heat gain, counteracting the loss of heat to the environment. The insulation value of the skin can be augmented by wrapping the body in furs, skins or fleece appropriated from other animals, or in material woven from plant fibre. Clothing enhances the thermal performance of the skin by retaining heat and moisture, enveloping the body in a miniature version of the tropical climate to which it adapted. In extremes of temperature, buildings can provide another layer of protection, augmenting the function of skin and clothing to provide a stable space suitable for human habitation.

As well as helping to avoid climatic extremes, the physical barrier of walls, floor and roof help to guard against insects and predators, providing basic levels of health and safety for the occupants. Buildings also provide shelter for fires, protecting them from the wind and rain, and enabling their heat to be retained for warmth. But fire in an enclosed space is potentially hazardous, not only because of the risk of burning down the building, but because of the competition for oxygen. To avoid suffocation, a regular flow of air is needed, allowing occupants to breathe and helping the fire to burn.

Buildings provide protection, but this alone is insufficient to explain the rich diversity of primitive and vernacular styles that characterize early human settlements. Aspects such as available materials and techniques of construction will affect the way buildings are made, and so too will spiritual beliefs and practices. But one of the major determinants of building form is the patterns of social interaction that occur within them. Once buildings provide basic shelter from the elements, they set up barriers that affect physical interaction between members of a group. By filtering light and sound, walls help to determine levels of interaction and privacy, affecting what can be seen and heard.[8]

The need to control light and sound is not a matter of biological necessity; rarely do these forms of energy occur at levels that are threatening to life. Instead, the visual and acoustic requirements of built space are determined by the social activities and events that take place within buildings. Thus while the performance of buildings begins with the need for shelter, the requirements for spatial and environmental control have evolved in concert with the increasing complexity of social and cultural institutions, such as those for education, health, justice, recreation and commerce. Each of these has different needs for the interaction of people within, and for the interaction between internal space and the external environment.

In almost every culture, the making of built space may be motivated by the need for shelter, but is ultimately transformed into an opportunity to explore the expressive potential of building. The construction of elaborate temples, monuments and palaces can be seen, on the one hand, as an expression of complex social or religious belief systems and, on the other, as a sheer celebration of human ingenuity. Similarly, the environmental conditions within such buildings are poetic as much as pragmatic: the exquisite shaft of light falling through the oculus of the Pantheon, for example, is a symbol of divine presence, not an aid to vision.

Passive and active systems

Buildings affect the flow of energy, of heat, light and sound necessary for people to stay alive and to engage in social activities. They also affect the air we need to breathe and the water we need to drink and wash, and the way they, too, play a part in social space. Most of the heat and light in and around buildings comes initially from the sun; even the energy from fossil fuels is from sunlight captured by plants millions of years ago. And while burning fossil fuels can release that stored energy in controlled form, occasionally fires escape control to damage or destroy buildings and to threaten the lives of their inhabitants. Sun, heat, light, sound, air, water and fire are the forms of energy and matter that flow in and around buildings, making them habitable and bringing them to life. In each of the chapters that follow, these factors will be addressed in terms of their impact upon built space. An understanding of how each of these is modified by building fabric is necessary for any architect, to make spaces habitable and to celebrate the possibilities for sensory experience.

Energy, air and water all flow in and around buildings in ways that are affected by their design: by their size, shape and orientation, by the choice of materials used for walls, floors and roof, and by openings in the form of doors and windows. A roof provides protection from overhead sun or rain falling from the sky; walls protect from lateral intrusion of wind or wildlife; floors seal off the moisture or dirt that can penetrate from below; a window can allow sunlight to enter a room. This is known as *passive* environmental control, because the flow of energy, air and water is affected without parts of the building having to move. Each building element will have a different degree of permeability, depending on materials and construction techniques. Often these elements are built up of several layers to improve performance, which can be enhanced by the quality of jointing and details. Alternatively, spaces can work together to have a cumulative effect, such as a walled garden or courtyard that can act as an intermediate space between the interior and the surrounding environment.

Another way of modifying the internal environment is to bypass the building fabric, using building services to control the flow of energy, air and water. Services are often described as *active* environmental systems, to contrast with the passive effects of building fabric in modifying the environment. Active means can include air-conditioning to control the temperature and flow of air, or plumbing to control the flow of water, or electricity to give light or heat at any time of the day or night. By consuming energy, systems such as artificial lighting and ducted air-conditioning can be used to precisely regulate internal environments without needing to rely on external conditions. The active control of environmental conditions usually requires a network of pipes or wires built into the building, along with appliances or fittings that convert or deliver the energy, air or water in a controlled manner to the point where it is used. Active forms of control also require ongoing maintenance, as well as access to resources such as water supply or energy to enable them to keep functioning.

The use of building services has transformed the types of environment that can be created within buildings, which has in turn transformed the types of building that can be built. High-rise office buildings, for example, rely on the structural innovations of steel framing or reinforcing in order to stand up, but rely equally on elevators to lift people to the upper floors, as well as plumbing to bring down their waste without a journey to ground level each time nature calls. Throughout the 20th century, buildings were increasingly designed to be reliant upon active forms of environmental control.[9] Now, in the early stages of the 21st century, there is a recognition of the need to create sustainable architecture, to reduce our reliance on active systems and to return to passive means of environmental control. However, this is not easy to do when actively serviced buildings set the standard for expected levels of comfort and control, and also affect the economic measures that determine the viability of any new development.

Architecture, science and technology

The flow of energy, air and water in buildings, whether through active or passive systems, has often been studied using scientific methods, especially those borrowed from fields such as physics and chemistry. Because of the physical behaviour of buildings in their environment, scientific techniques and ideas have long been seen as significant to the way buildings are analysed, designed and constructed.[10] A basic understanding of building performance, an essential part of every architectural curriculum, was taught for much of the 20th century within the field of *architectural science*. Several texts in this area are still available, including *Introduction to Architectural Science* by Steven Szokolay.[11] Other approaches have been to include aspects of building performance within the broader field of *climate-responsive design* or *sustainable architecture*.

The interest in sustainability shows how physical performance must be understood within particular social and cultural contexts in which buildings are used and interpreted. The flow of energy, air and water is affected as much by the people who use buildings as by the buildings themselves, by their attitudes and behaviour about suitable levels of comfort, or privacy, or resource and energy use. In describing the social and cultural aspects of building performance, it is helpful to understand architecture as a form of technology, bringing together the 'low-tech' influence of passive systems with the 'high-tech' features of active systems of environmental control.[12] Although the word *technology* tends to describe only the latest innovations, such as electronic gadgets, in a broader sense it describes any of the products of scientific knowledge that are used to extend human physical or sensory capabilities. Thus even the most primitive form of dwelling, by extending the range of climates that people can tolerate, can be considered a form of technology that augments or complements the human body.

In recent years, there has been a growing interest in the study of the social contexts in which technology evolves. In a field known as the *social construction of technical systems*, authors describe the way artefacts are not simply invented, but are developed and adopted within social networks that help shape their form and manner of use.[13] As new forms of technology become widespread, they are affected by the various interests and influences of manufacturers, users and regulatory authorities until they reach stable and socially accepted forms.

Many of the artefacts studied within this field are those commonly used in the built environment, from light bulbs, to household appliances, to electricity networks.[14] Even when other forms of technology are studied, their impact on built space can clearly be seen. A study of tram cars in America, for example, describes how the perception of increased danger led parents to stop children from playing in the street and relocate them to rear yards. Although this improved safety, it also reduced social interaction as households shifted focus away from the public realm and toward the private space of the garden.[15]

A range of new technologies served to shape the built environment from the late 19th century, especially those resulting from the reticulated supplies of electricity and water. These services led to new standards of cleanliness by making washing easier and reducing the need for fire inside the home.[16] Prior to the harnessing of electricity, heat or light could only be created by burning wood, candles, coal or gas. Not only did these various fuels need to be transported to the site of use, but the residues left from their burning, such as smoke, soot or ashes, needed to be removed. This placed limits on the amount of fuel that could be consumed, as the effects of combustion impacted directly on the users. With electrical appliances, however, there is no limit to the amount of light or heat that can be used, and no local impact in terms of waste. But as we have now discovered, the pollution exhausted at power stations many miles away has a different impact, as greenhouse gases contribute to global warming.

Whether low-tech or high-tech, architecture is profoundly technological. Using available techniques of fabrication and assembly, architects give order to the processes and products of the building industry in ways that reflect social and cultural needs and values. Like Lewis Mumford's history of civilization told through stages of technological development,[17] the history of architecture can be written in terms of the changing attitudes to technology: from the remarkable structures of the gothic cathedrals,[18] or the domes of the Renaissance,[19] to the innovations of steel and glass by the Chicago School, the German Werkbund and the Bauhaus, to the celebration of technology by the Futurists in Italy, the Metabolists in Japan or the High-Tech movement in Britain.

Innovations in materials and construction techniques have an impact on the form and appearance of any building, and thus affect the types of space that can be created and the experiential qualities that result. The relation between structure and enclosure, for example, gives rise to possibilities of fenestration, which in turn determines the quality of light for the building interior. Construction and fenestration type will also affect the thermal performance, while geometry and materials will affect the acoustic qualities of any space.

Along with changes to the passive environment, emergent technologies have transformed practices of active environmental control.[20] In the past century, active systems, especially air-conditioning and artificial lighting, have had a significant impact on the built environment. That impact was first described by Reyner Banham several decades ago, in his book *The Architecture of the Well-tempered Environment*.[21] The technology of environmental control, according to Banham, was as much a part of the history of architecture as technologies of construction. Active systems affect the way buildings are built, the way they are used and experienced by the occupants and, ultimately, the possibilities for architectural expression or style.

Active systems still have a significant effect on architecture, despite the growing concern over the damage to the environment caused by burning fossil fuels. Today, the case for sustainable architecture usually begins with the need to reduce reliance on active systems. Dean Hawkes, for example, argues against the 'exclusive' mode of sealed, artificial environments, in favour of a 'selective' mode of using building fabric to filter or select desired aspects of the ambient environment.[22] Other authors have described the way architectural styles are increasingly influenced by the expression of passive systems of environmental control.[23]

Comfort and sensory experience

Whether achieved using passive or active systems, the technology of architecture is intended to create spaces suitable for human habitation, providing a level of comfort and amenity that enables the various activities within buildings to be undertaken. In its original form, comfort meant alleviating pain and fatigue by strengthening or fortifying. But, as Siegfried Giedion observed, comfort became tied to a sense of domestic ease and pleasure afforded by the newly available furnishings and domestic appliances manufactured in the 19th century.[24] Comfort can foster human achievement by reinvigorating the body after the demands of work.[25] Or comfort can result from machines performing work on our behalf, using energy from fossil fuels to replace the labour of human hands.[26] Alternatively, comfort can be seen in terms of remaking the world in a way that is sensitive to human suffering, and reducing the effort required to meet the demands of the body.[27] This allows attention to be focused away from the body and outward into the world at ever-increasing distances.[28]

Unfortunately, the use of active systems has increased expectations about levels of comfort, leading to increased use of resources of energy, water and materials needed to make buildings comfortable. This is problematic for several reasons. First, once patterns of use have increased, it is difficult to lower them again. Second, the resources used in providing comfort are used indirectly, the result of 'inconspicuous consumption' where it is difficult to know how much people are using, or difficult for people to know how much is being used on their behalf.[29] Third, it may be possible to provide too much comfort, reducing the body's ability to change or adapt to its environment when it becomes necessary.

For example, active systems save us from the task of thermal regulation and it is now possible to enjoy ideal comfort conditions in the home, at work and in the car. The result, while comfortable, can lead to a sense of monotony or routine, dulling the senses from a lack of variation. To reinvigorate themselves, people often take vacations in diverse climates, such as the heat of a beach resort or the cool of mountains or snowfields. Sometimes the enjoyment of thermal extremes is more frequent, as with Japanese or Turkish baths or Finnish saunas. And when saunas involve alternating between the heat inside and the cold water or even snow outside, the aim appears to be more than just avoiding the weather. It may be that this challenging sensory experience – the thermal equivalent of extreme sports, perhaps – is a way to exercise the body's thermal response mechanism, improving the ability to withstand the variation encountered in daily life.

Active systems allow precise control of various aspects of the environment, allowing lighting or temperature to be set at particular levels regardless of variation outside. The only problem is deciding what level is appropriate. When a small number of people are in a space, the level can be adjusted manually. But when a large number of people share the same space, it is impossible to please everyone, so the aim is usually to find a setting at which the smallest number of people are likely to be dissatisfied. This means that comfort settings are usually about the average of what everyone would want, if they were given the choice.

Constant levels are also used because the human body is better at sensing changes in the environment than it is at detecting absolute levels. In fact, when people are comfortable, they are likely to not even notice the environment around them. The reason for this is that the body is capable of adapting to changing levels, 'recalibrating' to suit the current environmental condition. Our eyes, for example, can adjust to the light when we move from a dark space into a well-lit room, or vice versa. The change is perceptible because of the need to wait a few moments while the adjustment occurs. Constantly alert to the need to make adjustments, the body is more sensitive to changes in environmental stimuli than to absolute levels.

Environmental diversity

Avoiding the need for adjustment can prevent distraction, but it can also lead to fatigue and make spaces feel sterile. It is possible that stress can result from too little stimulation as well as from too much and that a small degree of adjustment throughout the day is helpful to keep the senses active. Buildings can protect from extremes of environmental change, but by connecting with external space, can also allow us to enjoy the variation that occurs naturally throughout the day and year. When electric lighting was first introduced, its constancy was promoted as an advantage over the unreliable variability of daylight. Now the variation in daylight is considered an advantage, with subtle changes due to a passing cloud or the shadow of a tree moving in the breeze making us aware of external conditions without causing distraction. Garden views are particularly valued because of the ever-changing colours, sounds and smells that enable us to feel in touch with the natural environment.[30] Changes resulting from natural variation are also more likely to be tolerated by building users, with the flicker of a candle, for example, being far more pleasant than the flicker of a faulty light fitting.

The need for a degree of sensory stimulation highlights the importance of environmental diversity in architecture.[31] Much of that diversity can result from the formal and spatial inventiveness of architectural projects. It can also result from the way buildings are designed to embrace changes in the external environment and bring them into the building interior. In Shoei Yoh's Light Lattice House in Nagasaki (1980), for example, a fine grid of sunlight tracks its way across the interior throughout the course of the day. This will happen with most windows, but can be emphasized or highlighted by considering the pattern of sunlight on the ground or by marking its place on the floor.

Because architecture is three-dimensional, it allows the opportunity for both spatial and temporal change. As people move towards, into and around any building, they are able to experience its forms and spaces from different locations, enjoying diversity simply by changing position in relation to the building and its external environment. By actively creating diverse environmental conditions within the various internal spaces, an architect can create a rich sensory environment for inhabitants as they move from dark to light, from cool to warm, from open to enclosed, or from noisy to quiet spaces. Our senses can be brought alive by the experience of architecture.[32]

While inhabitants can create environmental diversity simply by moving throughout a building, they may also contribute to that diversity by adjusting windows, blinds or shades. One of the key tasks of the architect is to design for these adaptive opportunities, to anticipate the active involvement of the building users in controlling the internal environment by adjusting the level of interaction with the outdoor environment.[33] For those who are unable to adjust building elements, adaptation may arise from a variety of spatial opportunity, moving between rooms designed with different environmental conditions or intended to be used at different times of the day or year.

Through variation in form and space, material and use, temporal change and adaptation, architecture can provide a diversity of experience for all the senses. However, this diversity is often neglected because of a view that architecture is concerned mainly with the visual appearance of buildings. Juhani Pallasmaa has identified that the dominance of vision in our society has led to a lack of consideration of the other senses. He argues for the need to consider the rich complexity of human experience, by considering the effect of touch and taste, sound and smell, in architecture.[34]

The dominance of vision might be explained by the way it is able to substitute for other senses. In particular, many of the functions of touch are performed in advance by vision. Having passed through that necessary phase of childhood where everything must be handled or placed in the mouth, we evaluate the tactile qualities of an object mostly by sight. Whenever we see an object, we instinctively anticipate its size and shape in relation to our body; its weight and texture, even its temperature. We evaluate cleanliness by sight, possibly aided by smell, as we look for patterns or dirt on the surface. Only when we see something particularly inviting are we inclined to reach out, wary of the injunction 'Don't touch!' common in modern society. Vision also helps us to evaluate factors for which we have no dedicated sensory apparatus: we may have a sense of enclosure or privacy, of scale and orientation, of pattern and order.

This substitution of vision indicates how the senses can interact, working together to build our awareness of the environment around us. In extreme cases, this can lead to a condition known as *synaesthesia*, where sensations from one sense are experienced in place of another.[35] But even in everyday experience, the senses cross-connect: when we say a space is 'warm' we might refer to its temperature, its lighting and colours, its acoustics, or all of these at once.

While the traditional division into five senses, of sight, hearing, smell, touch and taste, is convenient for distinguishing parts of the body, it is not very helpful in describing the dimensions of sensory experience. Taste, for example, involves a complex mix of flavour, texture and temperature as food is held in the mouth while eating. The senses also support each other, working together to give complete, multi-sensory experience. The sound and smell of a fire, for example, and the hypnotic sight of the flames, are all integral to the experience of its warmth.[36] In a similar way, the experience of space involves all the senses working together, not only sight and sound, temperature and odour, but also a sense of the position and orientation of the body in relation to the size and shape of the space.

The flow of energy, air and water affects various aspects of architecture, from environment to function, from comfort to sensory experience. Yet, to consider any of these in isolation is to overlook the role of the architect in reconciling or coordinating the complex and often contradictory requirements for built space. These various dimensions of environmental control are part of a range of aspects that must be resolved, including client needs and site conditions, available materials and construction practices, planning requirements and building codes, as well as anticipated changes, adaptations or future uses. The design of environmental conditions in any building, whether achieved using passive or active means, are interconnected with these other aspects, forming an essential part of the social, technical and aesthetic dimensions of built space.

chapter one
SUN

I tried to draw my shadow once, but I couldn't ... My arm kept moving.

Steven Wright

American architect Louis Kahn once remarked that the sun never knew how great
it was until it struck the side of a building.[1] Conversely, we might say buildings are
never greater than when they are brought to life by the sun. The sun provides warmth
and light, and marks the time of the day as it moves through the sky. Even when the
sun is overhead, its position can be inferred from the shadows it casts on objects
around us, helping us to perceive their shape and texture. The varying intensity of
the sun at different points on the Earth's surface causes differences in climate, which
lead to differences in the buildings made to respond to that climate. The sun marks
the seasons as its daily path varies through an annual cycle from the heat of summer
to the cold of winter, and back again. The sun brings life and energy to plants through
photosynthesis, and in turn to the fauna that eat those plants, and each other, along
the food chain. The sun forms an essential part of our daily life as we alternately seek
refuge from its intensity or revel in its warmth.

In understanding the role of building fabric as an environmental filter, the first thing
that must be filtered is the sun. For the most part, buildings do not move, and by
remaining in place are exposed to the changing position of the sun through the day
and year. Buildings that let sun in may be too hot and bright in summer, while those
that keep sun out may be too cold and dark in winter. In general, we prefer buildings
to work counter to the prevailing conditions, letting in the scarce winter sun and
keeping out the abundant summer sun. But even better is to design spaces in which
it is possible to enjoy the sun at all times of the day and year, capturing its warmth or
protecting from its heat when appropriate. This requires detailed understanding of
the geometry of the sun, so that its position at different times of the day and year can
be anticipated and thus incorporated into the design of built space.

Solar geometry

The regular changes in the path of the sun have been understood for many centuries. It is now generally agreed that the monument at Stonehenge is aligned to mark the summer solstice in that region. But mapping the true path of the sun meant dislodging the Earth from its privileged place at the centre of the universe. Since the publication of *De Revolutionibus Orbium Coelestium* by Nicolaus Copernicus in 1527, the Western world has accepted that that the Earth revolves around the sun, and not the sun around the Earth. But while this model of the solar system is widely accepted, it appears to have diverted attention from the *apparent* path of the sun in the sky and its very real consequences for the changing environment throughout the day and year. Few people, apart perhaps from astronomers and enthusiastic architects, can actually describe the path of the sun through the sky with any accuracy.

Sun and seasons

Our experience of the rising and setting of the sun each day is caused by the rotation of the Earth around its own axis. Our experience of a year is in turn caused by the revolution of the Earth around the sun. But the *reason* we experience a circuit around the sun as a 'year' is because of the seasons, the cycling back and forth between warm

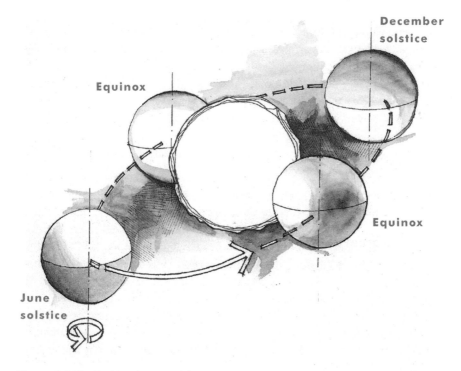

Figure 1.1 The Earth's orbit around the sun

summer and cold winter weather characteristic of temperate zones. What causes seasons to occur is the lack of alignment between the axis of rotation and the axis of revolution of the Earth – there is a slight 'tilt', known as the *declination*, between the two. The declination, currently an angle of about 23.5° (it wobbles back and forth in a cycle that lasts about 41,000 years), was probably caused by a large asteroid colliding with the Earth during its early formation.

Without the declination, there would be no seasons; the sun would rise and set along exactly the same path each day, and there would be nothing (other than a very subtle change in our view of the planets in the night sky) to mark the passing of the years. The declination causes seasons because it gives rise to changes in the path of the sun throughout the year, with the northern and southern hemispheres taking turns to tilt towards or away from the sun. To understand these changes and their impact on buildings, it is helpful to understand how they are experienced at different points on the Earth.

Annual sun path

Figure 1.1 shows the relation between the rotation of the Earth on its own axis and the revolution of the Earth around the sun. Drawing this with the axis of rotation vertical shows the December solstice (usually around the 22nd of the month), as the highest point reached by the Earth in its path around the sun, while the other solstice, occurring around 21 June, marks the lowest point of the same path. The equinoxes, occurring about 20 March and 22 September, mark the midpoints of the cycle, where the Earth appears to be level with the sun. The apparent path of the sun is affected by whether the Earth is at a low or high point in its journey around the sun. But how this is seen depends on where you are on the Earth's surface, with the curvature of the Earth causing viewers at different latitudes to see the same solar path from a different angle. This can be understood by imagining how the sun appears when standing at various latitudes.

Polar sun paths

For a building near the North Pole the path of the sun is always close to the horizon. At the June solstice, the rotation of the Earth makes the sun appear to travel in a low circle around the sky, just above the horizon, as this part of the Earth is tilted slightly towards the sun. The length of a day, 24 hours, is marked not by the rising and setting of the sun, but by the sun travelling one full circle around the sky. As the year progresses, the sun continues to circle around the sky, but its angle above the horizon gradually decreases (see Figure 1.2). By the time of the March equinox, the sun is circling level with the horizon, in a very drawn-out sunset. For the next six months, the sun continues to circle, but now it does so below the horizon, which is experienced as one long period of darkness and cold as this part of the Earth is tilted

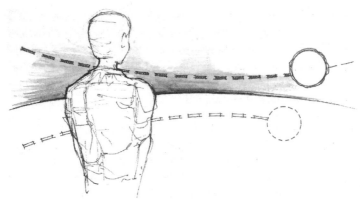

Figure 1.2 Polar sun paths

away from the sun. Thus the poles experience the most extreme form of 'seasons' – a long summer of sunshine followed a long winter of darkness – but there are no 'days' in the familiar sense of rising and setting sun.

Tropical sun paths

For a building at the equator, the sun is more intense because it appears higher in the sky. At the June solstice, the rotation of the Earth makes the sun move perpendicular to the horizon, in the northern part of the sky. It rises in the east, sets in the west, and in between follows a high path, with about 12 hours of light and 12 hours of darkness in any single day. As the Earth moves around the sun and reaches the March equinox, the sun still appears to rise and set, but moves gradually from the northern part of the sky to being directly overhead. For the next six months, the days remain unchanged, but the path of the sun gradually moves into the southern part of the sky, and then back again (see Figure 1.3). The amount of sun in any given day is effectively constant; there are no seasons caused by the changing length of the day or by the changing angle of the sun path above the horizon, just a constant rising and setting of the sun, with the day divided evenly between light and dark.

Temperate sun paths

Now imagine standing part way between the poles and the equator, in the *temperate* latitudes that contain most of the world's population. Here the path of the sun is part way between the long, drawn-out changes that occur at the poles and the constant cycling of days at the equator. There is still a rising and setting of the sun each day, but the days gradually get longer in summer and shorter in winter, giving rise to the seasons. This is because the path of the sun lifts above the horizon in summer, with more than half of it visible in any given day. In the northern hemisphere, this happens around the

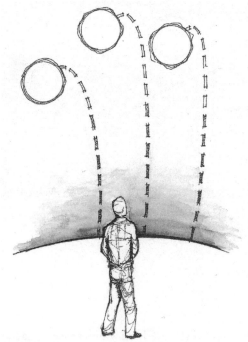

Figure 1.3 Tropical sun paths

June solstice, or summer, which is when the southern hemisphere is having shorter days, or winter. Six months later, at the December solstice, the northern hemisphere has shorter days, marking winter, while the southern hemisphere has longer days of summer (see Figure 1.4). In the temperate zones, the heat of summer results less from overhead sun than from the long, low light of the sun as it circles around the sky.

Sun angles

At any point on the Earth's surface, the path of the sun appears to change throughout the year by an amount equal to the declination, or 23.5°, either side of a midpoint that occurs during each equinox (see Figure 1.5). At the poles, where the latitude is 90°, the sun will be level with the horizon, or 0°, during the equinox, and then rise and fall above and below this point to a maximum of 23.5°. At the equator, where the latitude is 0°, the sun will be directly overhead, or 90°, at solar noon on the equinox, and will then vary by 23.5° either side of this point, to be at a maximum angle at solar noon of 66.5° in the northern sky for the June solstice or in the southern sky for the December solstice. For all points in between, the sun reaches an angle of 90° minus the latitude at solar noon during the equinox, and then varies by 23.5° either side of this for each solstice. The maximum sun angle at solar noon for summer and winter solstice can thus be found by the equation (90 – latitude ± 23.5°).

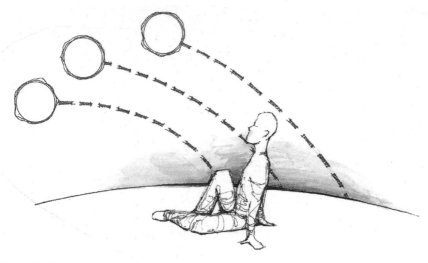

Figure 1.4 Temperate sun paths

Since the sun can only be directly overhead within a narrow band (± 23.5°) either side of the equator, the sun appears from the temperate zones to be in that part of the sky that is towards the equator. In the northern hemisphere, the equator is towards the south, so the sun, rising in the east and setting in the west, travels through the southern part of the sky.[2] Thus, in the northern hemisphere, south-facing windows will receive sun during the day, and north-facing windows will receive light from the sky, but not sun. In the southern hemisphere, the equator is towards the north; the sun still rises in the east and sets in the west, but here travels through the northern part of the sky. In the southern hemisphere, sun will enter through north-facing windows during the day, while south-facing windows receive light, but not sun. To avoid this confusion, the term *equator-facing* window is sometimes used to describe a window that will receive sun for much of the day, whether in the northern or summer hemisphere.

The cycles that mark the path of the sun throughout the year are also the source of the imaginary lines that define the globe. The Tropic of Cancer (23.5° north) is the latitude where the noon sun is directly overhead at noon during the June solstice; the Tropic of Capricorn (23.5° south), is where the noon sun is directly overhead at noon during the December solstice. The Arctic and Antarctic circles (23.5° away from the poles) are the latitudes where each solstice is experienced as either one day of full darkness or one day of full sunshine. These lines also mark the boundaries between the broad climate zones: the tropics, occurring between the Tropic of Cancer and the Tropic of Capricorn at latitudes below 23.5°; the polar regions, occurring within the Arctic and Antarctic circles, at latitudes above 66.5°; and the temperate zones, at all other latitudes between the tropics and the polar circles.

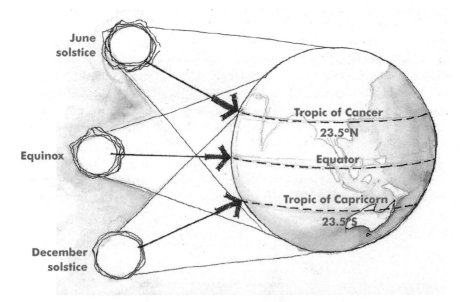

Figure 1.5 Apparent change in the sun's position throughout the year

Climate zones

The shifting of the sun back and forth each year from its midpoint above the equator helps to disperse the heat of the sun out towards the poles, assisted by atmospheric and oceanic currents. But even with this dispersal, the tropics remain the warmest part of the Earth, and the poles the coolest. As the angle of the sun gets lower in the sky, the intensity of radiation is reduced, since it loses more heat by travelling obliquely through the atmosphere and must also spread out over a greater area when it reaches the Earth's surface. The intensity of the sun at the tropics also creates warmer air and greater rates of evaporation, which in turn creates the higher levels of rainfall in this region. Thus the different path of the sun leads to the broad climatic differences between tropical and temperate regions. Actual climate conditions are also determined by a complex interplay of various factors, including atmospheric patterns, relation to land mass and water bodies, topography and vegetation. Within a given climate zone, a particular building will also be affected by microclimate conditions, depending on whether the site slopes towards or away from the sun or prevailing wind, whether it is shaded by adjacent buildings or vegetation, and so on. But in broad terms, buildings at a given latitude will share the same climate conditions and apparent sun path, and are thus likely to adopt similar strategies for dealing with the sun. These will be examined under the four broad areas of tropical (hot humid), arid (hot dry), warm temperate and cool temperate climate zones.

Tropical (hot humid): The tree-house

In tropical areas, the intense heat of the sun leads to high average temperatures, and also high humidity levels, which make the air feel hotter by limiting evaporation from the skin. Kuala Lumpur, for example, has a mean daily temperature of 28°C and an average humidity level of 80 per cent. The humidity also creates cloud cover, which can keep heat in at night because of the 'blanket' effect. In this climate, vernacular buildings create thermal comfort using overhead shading to protect from the heat of the sun, and open or permeable walls to maximize ventilation to promote cooling (see Figure 1.6). Roof overhang to the north and south can provide shading, while a narrow building plan with small east and west facades can avoid heat build-up in morning and afternoon. The high angle of the sun as it moves between northern and

Figure 1.6 Traditional tropical house

southern sky means that large overhangs are not essential, but these can be used to provide shaded spaces adjacent to the building as well as protection from monsoonal rain. Ventilated roof spaces allow heat to escape. Lightweight walls are used to prevent the build-up of heat. The use of louvres allows ventilation while limiting transmission of light, thus reducing the sense of being exposed to the heat of the sun. Enclosure is avoided, since this only serves to increase temperature and humidity by trapping the heat and moisture emitted by the occupants. Buildings are often raised on stilts to keep occupants away from the heat built up in the ground and to enable heat loss through floor as well as walls. The traditional Malaysian house is indicative of this approach, while more recent projects in that country by architects such as Jimmy CS Lim use similar strategies to provide thermal comfort and connect to local traditions.

Arid (hot dry): House as oasis

Arid climate zones tend to share the intense heat of the tropics, but with reduced humidity, because they are dominated by air masses formed over continents rather than oceans. The lack of humidity in the form of cloud cover allows heat to escape at night. This is known as *night sky radiation* and means that hot days are offset by cool nights. Cities such as Cairo (30° north) have high average maximum temperatures (28°C) and low average minimum temperatures (16°C). In contrast to tropical zones, humidity is much lower, around 47 per cent, which facilitates evaporative cooling whenever water is available. Traditional building types in arid climates tend to have solid external walls, turning away from the heat of the sun, opening instead to an internal courtyard space (see Figure 1.7). The courtyard may be shaded by a canopy as well as by the building that surrounds it. Where possible, the courtyard will include a fountain or plants to add water vapour to the air and make it feel cooler. Heavy walls encourage *time lag*, slowing the transmission of heat by half a day so that the walls can absorb heat from the sun and then release it to the interior during the cool of night. Small windows limit the amount of light and heat that can enter, with heavy walls allowing deep reveals to further shade the interior. Light external colours reflect the heat of the sun, while shading and small windows make interiors dark, enhancing the sense of shade. Walls facing into the courtyard are often lightweight open lattice or screen, allowing air to enter from the cooler courtyard space. Buildings are built on the ground to offset the extremes of daytime and night-time temperatures, and *malkafs*, or windcatchers, are sometimes used to channel warm air down over water or through the ground to cool it down before entering the building at low level. These vernacular typologies have been reinterpreted by architects such as Hassan Fathy in Egypt and Luis Barragan in Mexico and, more recently, adapted to suit the Arizona desert by American architect Rick Joy.

Figure 1.7 Heavy walls used in arid zones

Warm temperate: The outdoor room

In the broad range of temperate zones, spanning the latitudes from the tropics of Cancer and Capricorn to the Arctic and Antarctic circles, changes in the sun path throughout the year are experienced as seasons of summer and winter. In the areas closer to the tropics, known as warm temperate (or sometimes subtropical) zones, winters are mild, but shelter may be needed from the intense heat of the summer sun. Many of the strategies for keeping out heat are similar to those for tropical climates, such as narrow plans minimizing gains from morning and afternoon sun, and protection from overhead sun using ventilated or insulated roofs. However, buildings in this zone may include other strategies, such as large equator-facing windows with an overhang to block summer sun but allow winter sun to enter, preferably opening to an outdoor space that can be shaded in summer and sunny in winter (see Figure 1.8). They are also likely to be built on the ground to take advantage of the moderating effect of a lower ground temperature, with a slab possibly providing thermal mass to absorb heat from winter sun. Heavier masonry walls may be used in spaces away from the sun, such as north-facing bedrooms, to moderate cooler night or winter conditions.

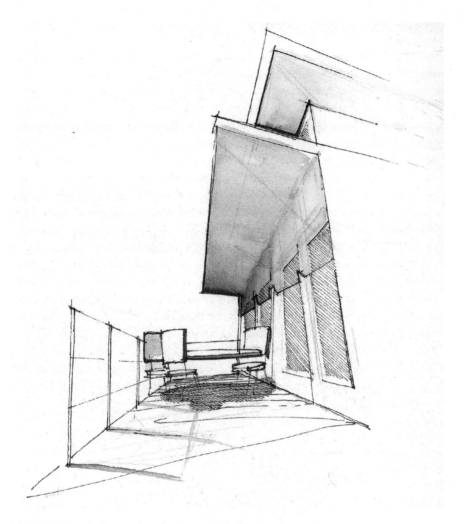

Figure 1.8 Sunshading and outdoor space

In general terms, the approach here is to interact with exterior space, with living areas, although under the main roof, having large windows or doors to connect to the outdoors. In the particularly benign climate that occurs for most of the year in warm temperate zones, buildings need only provide minimum protection, providing shelter from the heat of overhead sun and access to facilities for washing, sleeping and storing food. Most daily activities, including cooking, can take place outdoors. Donovan Hill architects, working in the warm temperate climate of Brisbane in Australia, describe the importance of the 'outdoor room', a large living space surrounded by other parts of the house, that can be closed off if necessary, but is mostly left open to the external conditions.

Cool temperate: House as hearth

As latitudes increase, temperate climates become cooler, with mild summer weather but more severe winters, when shelter is needed from the cold and rain. The source of winter heat, such as a fireplace, is often the centre or focus of the whole house, originally used for cooking as well as warmth, and with bedrooms above to catch the heat as it rises. Walls of high thermal mass store available heat or are heavily insulated to prevent heat loss to the exterior. A more compact plan form makes the house easier to heat and reduces the surface area of the walls through which heat may be lost. Smaller windows also minimize heat loss and may be double-glazed, as well as sealed against draughts and rain. Strategies for collecting winter sun, such as large, equator-facing windows or even a sunroom, may also be included, with living spaces opening to outdoor areas in warmer summer months. Spaces such as kitchens or bathrooms may be located to the east to let in morning sun, and a verandah or deck may be built to the west to capture the warmth of the late afternoon sun. In cool temperate climates, buildings may also 'huddle together' to conserve warmth, with terraced (row) houses conserving heat by sharing common walls, a fireplace and a staircase on either side allowing people and heat to rise up through the building to meet at upper levels (see Figure 1.9). The idea of house as hearth is most easily associated with the work of American architect Frank Lloyd Wright, from the early work in Oak Park in Chicago to the later Usonian houses. But the combination of fire and sun is a common theme in many houses, as architects prepare for the changing cycles from winter to summer and back again.

Figure 1.9 Terraced housing suitable for cool temperate zones

Sunshading

These basic typologies reflect the sort of environmental responses typical of vernacular architecture, adapted to climate through necessity using available materials.[3] They also reflect the adaptation of vernacular traditions by architects aiming to apply these important cultural and environmental strategies.[4] These traditions, evident in the four basic types above, show how climate-responsive strategies of building form, construction, orientation, fenestration and sunshading afford protection from extremes while taking advantage of favourable conditions. The key element in all of these strategies is an understanding of solar geometry so that the sun's heat can be captured or blocked at different times of the day and year.

The main aim of sunshading is to intercept the heat of the sun before it reaches other building elements, especially windows, thus allowing heat to remain outside the building instead of passing through to the interior. Since heat from the sun can reach up to 1000 watts per square metre, the heat gain through unshaded windows can be extreme. Trying to stop the heat after it has passed through the glass – by using curtains, for example – is far less effective than external shading, since the sunlight is absorbed as heat, which then becomes trapped by the glass. But sunshading devices can also be designed so that they block sun during the hotter parts of the year and yet let it in during the cooler winter months. This is easy to do for equator-facing windows, but harder for other elevations.

The design of windows and sunshading devices depends primarily on what the space is to be used for as well as on the general climate conditions, since light and heat are a matter of function as well as comfort. Also, the path of the sun at a particular latitude must be considered in connection with local conditions: windows will not capture south sun if blocked by another building, and blocking west sun makes no sense if that is where the best views happen to be. The need to reconcile 'global' conditions such as climate and sun path with local conditions of site and aspect gives rich opportunities for the design of fenestration and sunshading. It also explains why buildings *should* be designed, and not mass-produced without concern for their context.

Solar charts

While it is easy to construct overhangs or canopies above windows to protect them from the sun, the design of an accurate sunshading device, letting sun in during winter and keeping it out during summer, requires a detailed knowledge of solar geometry, as represented in the solar chart. The chart is like a map of the whole sky, with the position of the sun at different times of the day and year projected down onto the map with the horizon at the edge and viewer (*you are here*) at the centre. Any projection of the dome of the sky down to the circular map will always involve some distortion – like trying to squash half an orange down onto a flat surface. (The same problem confronts cartographers, who aim to draw the features of the Earth on flat sheets of paper.) The two main approaches are either to project vertically down from each point in the sky (*orthographic projection*) or to angle the projections inward so that each part of the sky gets a more 'even' share of the area of the circle (*stereographic projection*). The first approach tends to preserve the shape of the sun path, but makes the chart hard to read as the lines get closer together towards the edges (see Figure 1.10). The second approach makes the chart easier to read, but makes the paths of the sun for summer and winter no longer appear parallel (see Figure 1.11).

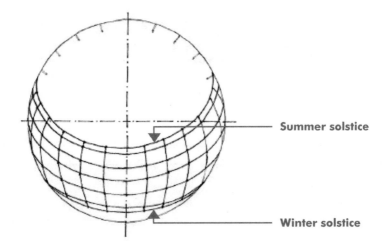

Figure 1.10 Orthographic solar chart

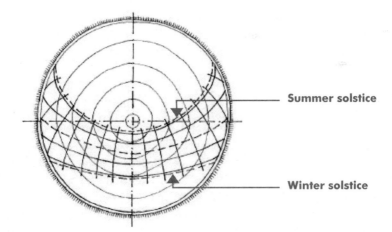

Figure 1.11 Stereographic solar chart

Shadow projection

The solar chart can be used to predict the availability of sunshine and shade throughout the day and year for various spaces at design stage.[5] The use of manual techniques has been superseded by computer modelling, with most three-dimensional software programs allowing latitudes to be specified, giving accurate shadow projection for any design. Dedicated programs for shadow projection, such as Ecotect, can also be used. Many planning authorities, such as local councils, require plans showing shadow projection to ensure that new buildings will not have adverse effects on the amenity of existing residents by blocking sunlight to windows or outdoor spaces. Designers can also use shadow projection to develop building envelope studies based on shadows cast by adjacent buildings.[6] Shadows are also essential for illustration techniques, revealing patterns of light and shade in internal spaces when applied to sections, and giving depth and contrast to elevations and other drawings. The construction of accurate shadows in orthographic drawings, known as *sciagraphy*, was once an essential part of any architectural curriculum.

Daylight saving

The solar chart shows that the length of the day gets longer in summer, but does so symmetrically around the highest point at noon (solar time). In other words, as the days get longer, the sun sets later in the evening, but also comes up earlier in the morning. With *daylight saving*, when clock time is moved forward one hour during summer, everyone's daily activities occur at the same time according to the clock, but an hour earlier in relation to the rising and the setting of the sun. This enables the longer hours of daylight to be 'saved' for the end of the day, when they are more likely to be enjoyed. The reason daylight saving was introduced, however, was not simply to make evenings more enjoyable, but also to reduce energy consumption, allowing civic services such as street lights to be turned on later at night, with the added benefit of reducing private energy use.

Sunshade design

The design of a shade for an equator-facing window can be easily determined from the maximum sun angles at solar noon in summer and winter, as shown in Figure 1.12. By projecting a line up from the bottom of the window at the maximum summer sun angle, and then another line out from the top of the window at the maximum winter sun angle, the point at which they cross can be used to mark the edge of a shading device. A shading device built at this point will have the remarkable property of blocking the sun out completely at the summer solstice, then progressively letting more in each day until the winter solstice, when the whole window is bathed in sun (weather permitting!).[7] To prevent sun coming in through the window as it moves up to and then beyond the midpoint at solar noon, the shade should be extended past the edge of the window on each side by about half the height of the window or turned down along the edge of the window to create a shading 'hood'. If equator-facing windows are used in a space containing a concrete floor, the concrete will act as a thermal mass, storing the heat from the sun in winter and re-radiating it into the room. This will help to reduce heating costs and allow the warmth of the sun coming through the windows to be enjoyed long after the sun has set.

These strategies are particularly suited to warm temperate climates where there is a need to keep out summer sun: for Rome, latitude 41.5° north, there is a maximum winter angle at solar noon of 25°, and a maximum summer angle at solar noon of 72°. For London, latitude 51.5° north, the angles would be winter 15° and summer 62°. These angles can be read from a solar chart or derived from the equation (90 – latitude ± 23.5°). It is also helpful to note that the geometry of the shade is independent of scale – it is possible to create a single shade for each window, or to use two, three or more shades stepped down the height of the window. With external louvres, each blade acts as a sunshade for the section of glass below it, keeping out summer sun and letting in winter sun. The resultant articulation of a single shade into 30 or so smaller shades also creates dramatic patterns of light and shade as the sun enters the space. For more northerly latitudes, the need to block summer sun is less important than the need to capture winter sun, which can best be done using sun spaces (see below).

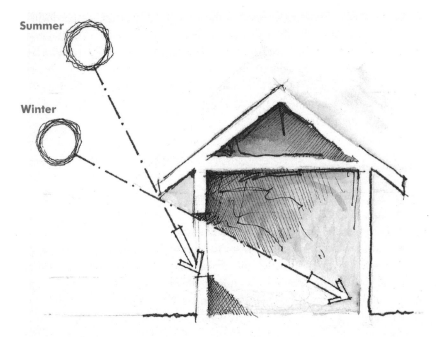

Figure 1.12 'Ideal' shading for an equator-facing window

Controlling the sun for equator-facing windows is made easy by the difference between the high summer sun and the low winter sun. But the design of shading for east and west windows is less straightforward. The sun always rises in the east, but does so from due east only at the equinox. In summer and winter, the sun rises either slightly north or slightly south of due east, depending on the hemisphere. Similarly, the sun always sets in the west, but alternates between north and south of due west depending on the season. It is possible to design sunshades that angle towards the winter sun path and away from the summer sun path in each direction, but these tend to be vertical and thus block out view. The usual strategy is simply to allow some sun to enter through east windows, especially into spaces that are used in the morning, such as kitchens or bathrooms, but to block out afternoon sun from the west using small windows, deep verandah-type spaces or deciduous trees. One final option for dealing with solar radiation is to avoid sunshading altogether and to utilize low-emissivity (*low-E*) glass designed to admit reduced amounts of light and heat from the sun to the building interior. The various forms of glass currently available are extremely effective at reducing heat while maintaining good colour rendering, but since they work during winter as well as summer, should only be considered as part of an overall strategy for solar design.

Sun spaces

As well as considering the solar gain for various windows, it is also possible to allocate a dedicated *sun space* in the form of a fully glazed room oriented towards the sun. Although the sun space itself may be uncomfortably hot, the warm air can be distributed to adjacent rooms using fans. A similar effect can be achieved using a Trombe wall, consisting of dark-painted masonry behind a layer of glass, capturing the heat of the sun and converting it into warm air that can be distributed through the rest of the house. However, using the equator-facing facade principally for heat gain can limit the interaction between inside and outside spaces in that direction. Sun spaces and Trombe walls are typically used in 'solar' houses, often combined with other devices such as solar hot-water systems and photovoltaic cells to meet the needs for heat and power using the renewable energy of the sun.[8]

Designing for sun and shade is perhaps the most fundamental of architectural decisions. It affects decisions about massing and orientation, about windows and sunshading devices, that determine the quality and character of any building and the spaces contained within. But design for solar access must also be integrated with other issues, especially the relation to the public space of the street or to available views, which are largely predetermined by planning issues such as site geometry and adjacent buildings. The need to conform to existing settlement patterns, whether presumed or mandated by planning regulations, will often lead to less than ideal orientation for solar access. One of the main tasks for any architect is to design for sunlight on awkward or difficult sites, resolving conflicts between orientation and privacy or security on external walls, or by bringing light down from above through atria or void spaces.

chapter two

HEAT

I have a microwave fireplace in my house. The other night
I laid down in front of the fire for the evening in two minutes.

Steven Wright

Buildings have become an integral part of human strategies for dealing with temperature change. When people feel too hot or too cold, they engage in a range of activities designed to speed up or slow down the rate of heat loss to their environment. When there is too much heat, people will seek shaded spaces, remove outer layers of clothing, immerse themselves in water or consume cool food or drinks. When there is too little heat, people may seek sunny spaces, increase layers of clothing, immerse themselves in warm water or consume warm food or drinks. Buildings can insulate and protect inhabitants from extremes of heat, wind, rain or sun, but can also provide a range of facilities for thermal control, including storage for clothes and blankets, facilities such as baths, showers or swimming pools, and appliances for heating and cooling food, water or air. A kitchen, for example, is not simply a place for making food, but an essential part of thermal adaptation that allows heat to be directly added to or removed from the body. Few things provide warmth and comfort as much as the smell of food being cooked. Similarly, other spaces, such as bathrooms, living rooms, sunrooms, courtyards or gardens, play an essential role in providing variety and choice of thermal experience.[1] 'The experience of home,' writes Juhani Pallasmaa, 'is essentially an experience of intimate warmth.'[2]

Adding and removing heat

Strategies for dealing with heat in space involve either adding it and retaining it when there is not enough, or removing it and keeping it out when there is too much. This is because heat is an absolute property, meaning that there can be no negative value: what we experience as cold is simply an absence of heat.[3] Heat occurs because of the motion or vibration of atoms, molecules or electrons within any substance. The point at which there is no heat is the point where the molecules have stopped moving, which occurs at absolute zero, or –273°C.[4] Since the average temperature of the Earth is 15°C, or 288° above absolute zero, there is clearly a large amount of heat already in every building that is conveniently close to the temperature of the human body. But even though heat is an absolute quality, the experience of heat is never one-dimensional. Instead, the various physical properties of materials give rise to complex patterns of heat storage and flow in and through the objects around us, as well as in and through the body itself, leading to different sensations of heat and cold.

There are various strategies available for adding heat to or removing heat from a space. Heat can be added directly, by capturing the heat of the sun, by converting electricity or by creating fire by burning wood, coal, oil or gas. It can also be added indirectly, using air or water heated by a central fire or boiler and carried in ducts or pipes. Fewer options are available for removing heat, which must usually be done by using water or air to absorb heat and then transfer it to another, usually external, space (see Chapter 5). Adding heat to or removing it from any space is one problem, but keeping it in or keeping it out is another. Once heat has been added to a space, making it warmer than its surrounds, the heat will naturally flow back out and must then be replaced with yet more heat. If heat has been removed from a space, making it cooler than its surrounds, the heat will naturally flow back in, only to be removed again. Thus, along with any strategy for adding or removing heat, it is essential to consider the way in which building fabric resists the flow of heat from one side to another, from inside to out or from outside to in.

Heat flow: Conduction

The way in which heat moves into and out of buildings will be affected by the thermal behaviour of objects and materials in the space, which will in turn have an effect upon the way heat is experienced by the occupants. Heat can be transferred from one object to another as a result of conduction, convection or radiation. With conduction, heat moves directly from one object to another as vibration is passed between atoms, molecules or electrons. The rate of flow depends on the temperature difference between the two objects; the greater the difference, the more rapid the heat flow. But it also depends on how quickly the heat can get from one part of an object to another, known as its *conductivity* (k). In general, conductivity is related to density, with more closely packed molecules able to transfer heat more rapidly. A material with low conductivity will resist the heat flow and is thus said to have a high *resistivity* (r). When touching a highly conductive material, such as metal, heat will move quickly towards or away from the rest of the object to the point of contact with the body, making the temperature difference seem greater. For materials that resist the flow of heat, the part of the object that touches the body will quickly reach the same temperature, even if the rest of the object does not. Materials such as timber or fabric will reduce the perception of temperature difference by preventing the flow of heat to or from the body.

Heat flow: Convection

The process of convection begins when heat is conducted between objects and the air that surrounds them. Air is in fact highly resistive to the flow of heat, provided it can be kept still. However, since air is a fluid, it can move heat from one place to another, not by passing it from molecule to molecule, but because the warm air can move away and be replaced by cooler air. Convection allows heat to move between objects at a distance, carried by the air between them. Convection can also be important for assisting heat loss through evaporation, as it removes moisture transferred from the skin to the air, allowing it to be replaced by dryer air that can then absorb further moisture. Evaporation cools because heat is absorbed by moisture, which becomes suspended in the air as water vapour. Evaporation converts *sensible* heat – heat that is felt in the air – into *latent* heat, which is held by the water vapour and experienced only as humidity.

Heat flow: Radiation

Even without air in between, heat will pass from object to object by radiation, as electromagnetic waves are projected through space carrying heat from warm objects to cooler ones. Radiation passes through the air directly to the body, making the skin feel warm when exposed to the sun or to an open fire. Radiation also passes from the body to any objects around it at a lower temperature, which in most climates occurs constantly, providing an important means of releasing unwanted heat. Radiation is affected by the surface of an object, with black objects being very effective at radiating or absorbing heat, while highly reflective materials, such as reflective foil insulation, can be used to reduce radiant heat transfer.

The sensation of heat

Although the heat that is transmitted may be the same in each case, the different forms of heat exchange – conduction, convection and radiation – are all experienced in different ways. Further, since all three are likely to occur at any given time, a broad variety of experience is possible within the overall thermal environment. In order to maintain its metabolic processes, the body must produce heat and then dissipate unused heat to its surroundings. At any given time, the body will dissipate heat through a combination of conduction, convection, evaporation and radiation. Although forms of heat loss vary, the body loses a small percentage of heat – about 10–15 per cent – from exhaling warm, moist air. Of the rest, about half is lost through radiation and the remainder is lost through convection/evaporation to the air. Heat loss from conduction tends to be limited, since objects with which the body is in contact for long periods of time, especially furniture, are usually well insulated. For the most part, the human body is the warmest object in its environment and the processes of heat loss are imperceptible. What is noticed are changes in the rate of heat loss or heat gain, especially those that may require adjustment. Radiant heat from the sun makes us feel warm even before we have warmed up; a cold breeze can cause a chill even before any heat is lost. The body anticipates the new thermal conditions and adjusts accordingly.

Measuring heat

Because of the different forms of heat loss, the traditional strategy of measuring air temperature, known as *dry bulb temperature* (DBT), provides only a limited view of comfort conditions. Most efforts to quantify comfort also take account of *relative humidity* (RH), which will affect the rate of heat loss from evaporation, as well as considering air movement within a space. Using a 'globe' thermometer, it is also possible to measure *mean radiant temperature* (MRT), which is the average temperature of all the surfaces to which the body can radiate heat from a given point. Some measures take account of both air and radiant temperatures, while others, such as *corrected effective temperature*, incorporate humidity levels and air velocity as well. These different ways of measuring temperature recognize that we may be equally 'comfortable' in different thermal conditions. For example, the same air temperature may seem different on a spring day, with the cooler ground being warmed by sunshine, than it would in the autumn, with the warmer ground being cooled by rain.

Heat from fire

From the very beginning, buildings have been heated by fire. One of the most fundamental roles of any building is to give protection from the elements, providing a place that is warm and dry by enclosing the heat of a fire within. Buildings have always depended on fire to be kept warm, dry and light; but so too does fire depend on buildings for protection from the wind and rain. 'Whoever desired fire,' writes Steven Pyne, 'would have to feed it, breed it, train it, shelter it – to sustain it in an artificial environment in which the desired forms of fire could thrive. In domesticating fire, however, humanity had to begin domesticating itself.'[5] Fire gives more than just heat and light; it can be used for cooking, extending the range of foods that can be eaten; it can be used for firing clays and smelting metals, transforming minerals into tools and materials for building. The embodied energy of brick and concrete, glass and steel, is needed mostly to produce fires hot enough to change the physical or chemical properties of raw materials, transforming them into a state that is suitable for construction.

Many early dwellings were simply a means of capturing the heat of a fire, with openings designed to remove smoke and depleted air. In warmer, Mediterranean climates, small charcoal braziers were usually sufficient for providing winter warmth. But in the colder climates of northern Europe, larger fires and fireplaces were needed, which were more easily built in a wall to one side rather than at the centre of a room (see Figure 2.1). The use of wood as a fuel source meant that people depended on the land for heat as well as for food; the larger the building, the larger and more numerous the fires needed to keep it warm, the more land needed to source firewood. By the time of the industrial revolution, large volumes of wood were needed for shipbuilding and for charcoal used in metal production, but coal had become readily available for use in domestic fires. Coal fires were smaller, but needed to burn at higher temperatures to be efficient. This eventually prompted a shift from open fires to enclosed, metal stoves, such as those invented by Benjamin Franklin in America (1740) or Count Rumford (Benjamin Thompson) while in Germany (1796). These *slow-combustion* stoves improved efficiency and reduced waste, allowing more heat to be enjoyed from the same amount of fuel (see Figure 2.2).

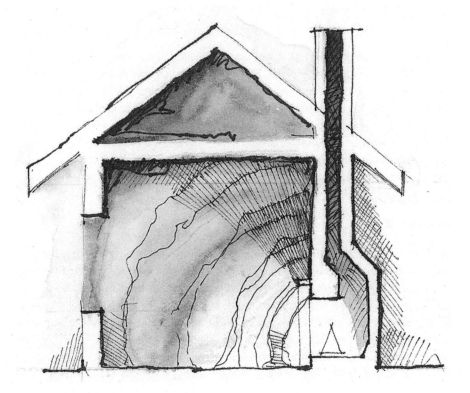

Figure 2.1 Fireplace providing radiant heat

Figure 2.2 Slow-combustion stove

Fireplaces and slow-combustion stoves are still used today, but often more for aesthetic reasons than for daily heating needs. Fire still has a primordial appeal, but the cost of wood and the effort of lighting fires and removing ashes makes daily use less appealing. Open fires are the least efficient heating source, since the need to vent the fire using a flue or chimney causes much of the heat to be lost to the outside. In areas where wood heaters are common, emissions can result in poor air quality, especially when fires and heaters are not properly maintained. One alternative is to install natural gas burners into fireplaces, which provide a cleaner fire without producing ash or other residue.

Heated air

Another way of heating buildings is to distribute air from a single heat source through ducts or flues into the rooms above. The Romans used a version of ducted heating, known as a *hypocaust*, to channel heat from a basement fire through raised floor cavities and up through flues built into walls. Ducted heating proved especially useful for distributing heat to various rooms in large public buildings. In 1844, a system of supplying warm air to cells was introduced at England's Pentonville prison, giving prisoners access to air without the disruptive influence of windows.[6] Today, warm air can be supplied either through a dedicated heating system or as the *reverse cycle* of an air-conditioning system. Because hot air rises, a dedicated system will usually require large floor ducts to deliver air, which are more easily built into lightweight

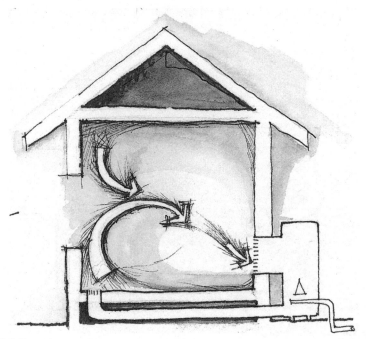

Figure 2.3 Ducted warm air

construction. When combined with a ducted air-conditioning system, the supply of air is usually from ceiling-mounted diffusers that require the air to be forced down to floor level before it rises back up again (see Figure 2.3). The greater air movement that results can detract from the sensation of warmth, since moving air is more likely to be associated with a breeze that can remove heat and increase evaporation from the skin.

Hydronic heating

A more effective system of heating emerged from the technologies developed as part of the industrial revolution. James Watt revolutionized steam power by adapting existing engines to make them more efficient. Watt also experimented with steam as a means to heat buildings and in 1784 ran pipes from the factory where steam engines were manufactured into the adjacent office.[7] Steam was forced under pressure out through the pipes and, as the heat dissipated, the steam condensed to water and returned to the boiler. Steam heating depended on the availability of metal pipes, which were sometimes made from discarded gun barrels. But the heat and pressure of the steam meant that these systems were often noisy and dangerous. Eventually, the development of electric pumps and thermostats made it possible to use hot water to distribute heat to radiant panels installed in each room.

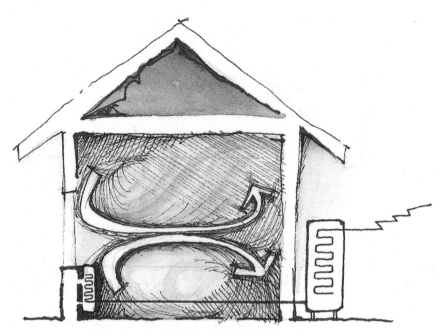

Figure 2.4 Hydronic heating

These *hydronic* systems proved extremely popular in both domestic and institutional buildings throughout the cool temperate climates of Europe and North America. These systems were also readily adapted to the new forms of energy supply, especially electricity and natural gas, which became available in the 20th century. The development of reinforced concrete construction also allowed pipes carrying hot water to be built into the floor, enabling the entire slab to act as a radiant heat source for the space. Frank Lloyd Wright was a proponent of in-slab heating, using it in several houses, such as the Jacobs house of 1937. Curiously, as Fernández-Galiano points out, Wright still included a fireplace in the Jacobs house, which was 'as functionally redundant as it was symbolically indispensable'.[8] By relegating combustion to a boiler usually located in a basement, hydronic systems allowed heat to be enjoyed without any of the adverse effects of flame and independent of the centralizing geometry of the fireplace. Because hydronic heating allowed heat to be delivered to any point in a room, it proved essential to achieving the abstract and homogeneous space that characterized Modernism.[9] The change to construction types, with more open space and increased levels of external glazing, made this type of heating indispensable.

Today, hydronic systems provide one of the most effective forms of heating for cool temperate climates. Although installation costs are high, the small-diameter pipes for carrying hot water can be easily built into the thickness of walls or floors, and the panels occupy minimal space. Panels act as a radiant heat source for nearby objects

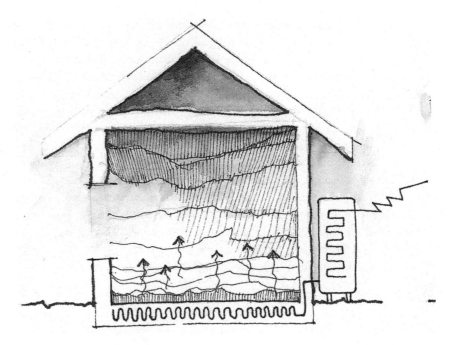

Figure 2.5 In-slab heating

and also warm the air, which then circulates through the space. Panels are typically located below window sills, where the rising warm air will meet with the air cooled by the window and be pushed back into the space (see Figure 2.4). (If located opposite a window, the heated air can cycle above the heads of occupants, only to be cooled by the window and then returned at floor level.) Hydronic systems operate silently and individual panels can be turned on or off. In-slab heating systems are less common, needing water pipes or electric coils to be tied to reinforcement during construction, and a greater depth of slab is required to give sufficient coverage to the pipes. By transforming the entire floor into a giant radiant heater, in-slab systems can operate at a lower temperature to achieve the same amount of heating and give a more even heat distribution across the room (see Figure 2.5). Because the slab takes a long time to heat up and cool down, this type of system is only suitable for situations where predictable patterns of weather and occupation are likely to occur.

Thermodynamics

The development of steam power not only transformed manufacturing processes, it also led to an improved scientific understanding of heat. While the English were making steam engines for use in the mills of Manchester, it was a Frenchman, Sadi Carnot, who first sought to describe their efficiency. Carnot, understanding the importance of heat, explained that the amount of work done by a steam engine depended on the difference in temperature between the hot and cold parts of the engine cycle (the boiler and the condenser). Because of the lower limit to temperature that occurs at absolute zero, any engine is always less than perfectly efficient, and in converting heat to mechanical motion, some heat is always lost or wasted. Carnot's analysis led to what are now known as the *laws of thermodynamics*. The first law of thermodynamics states that the amount of energy in a closed system is always constant, but can be changed from one type to another. The second law of thermodynamics is rather more difficult to describe, but loosely states that the order or organization of a system, including that of its energy, reduces every time an exchange of energy takes place. The *entropy*, or disorder of a system, increases, while the *exergy*, or usefulness of its energy, decreases. The result in most cases – especially with the engines studied by Carnot – is that 'waste' heat is produced. Now, heat is extremely useful when concentrated inside the cylinders of a combustion engine or turbine, but it is very difficult to gather up and reuse once it has spread to the surrounding environment through dissipation or friction.

Useful energy

The usefulness of energy, as described by the second law of thermodynamics, has important consequences for architecture. Although buildings are surrounded by energy in various forms, especially free energy from the sun, it is not always in usable form. Buildings depend upon regular supplies of usable energy, especially electricity, at all stages, from refinement of raw materials, to construction, to operation and maintenance, to adaptation, and eventually to demolition. Useful energy enables us to improve the 'usefulness' of the environment, by converting iron ore into steel, then steel into buildings, then buildings into habitable spaces that can be used as houses, offices, shops, hospitals, and so on. Fossil fuels are useful because the chemical energy they contain can be easily released by combustion and converted into heat, light or motion. Electricity is even more useful, because it can be converted directly into these other forms of energy without producing gas or waste at the point where it is used. It can also be used to manipulate data and images using electronic devices such as the computer. However, electricity is not itself an energy source, but simply a means of transferring it from one place to another. The energy used as electricity must first be created at power stations, which can be linked to renewable forms of energy such as solar, wind or water (hydro-electric) power, but which are more likely to be driven by non-renewable energy sources such as fossil fuels or nuclear energy.

Electric heating

Electricity can easily be transformed into heat, either with resistance heaters or with panel heaters that imitate the heat of hydronic systems. These heaters are highly efficient, converting electricity directly into heat without waste. However, generating that electricity is still likely to result in the production of waste, such as emissions from coal-fired power stations and heat losses along transmission lines. Transforming fossil fuels into electricity makes the energy more usable, but spreads the impact far beyond the place where it is used. Electricity helps to keep cities clean, but unless generated from renewable energy sources, contributes to the more extensive problem of greenhouse gases and their potential to cause climate change. Understanding the usefulness of energy is essential for creating 'sustainable' architecture.[10] Many strategies, such as improved efficiency of lighting or air-conditioning systems, better insulation, heat recovery and exchange, or cogeneration of heat and power, involve minimizing or reusing energy that would otherwise be lost as waste heat. Another aim is to minimize the use of high-grade energy, such as electricity, and utilize lower-grade energy, such as natural gas, especially for generating heat. Other strategies involve the use of renewable or ambient energy sources, capturing energy from the sun or wind that would otherwise go unused.

Urban heat island

Another problem is the accumulation of waste heat in urban environments. Heat absorbed by hard surfaces such as buildings and pavements, along with heat produced by machines, can result in city temperatures being several degrees higher than those of the surrounding countryside. This is known as the *urban heat island* effect, which can unfortunately lead to higher energy use, especially in summer when air-conditioners are used to cool building interiors. The effect can be lessened by increasing the amount of vegetation within a city, which provides cooling through evapotranspiration. In some cities, this is achieved by requiring all new developments to incorporate planted space equivalent to the *footprint* or ground floor area of any building, which often takes the form of a 'green roof'.

The sensation of heat

Whether delivered to a space by fire, air, water or electricity, heating is an active form of environmental control resulting principally from the conversion of chemical energy by combustion. Any heat introduced to a space will then be transferred to objects in the space – including the inhabitants – as well as to the air and to the building that surrounds the space. The active heating of any space will thus combine with the passive thermal performance of the building fabric, which will absorb the heat and slow down its dissipation to outside. The ability of a material to absorb heat is not the same as its ability to slow down the flow of heat, although these are interrelated. The former depends on a property known as the *specific heat capacity*, with each material used in construction having a different capacity to store heat, 'specific' to that material.[11] This is broadly related to density, with dense materials such as masonry and concrete able to store large amounts of heat. The latter depends on a property of each material known as its *resistivity*. Examples of each of these properties are given in Table 2.1.

Table 2.1 Conductivity, resistivity and specific heat capacity of common construction materials

Material	Conductivity (W/m°C)	Resistivity (m°C/W)	Specific heat capacity (J/kg°C)
Brick	0.89	1.12	840
Concrete	1.40	0.71	840
Fibre cement sheet	0.36	2.78	1050
Glass	1.10	0.91	840
Plasterboard	0.16	6.25	840
Timber	0.15	6.67	1200
Steel	50	0.02	480
Water	0.62	1.61	4187
Glass fibre batt	0.035	28.57	1000
Still air pockets	0.02	50.00	1006 (or 1300J/m³°C)

Estimating heating loss

Since materials differ in their capacity to resist the flow of heat, the rate of heat loss or gain between a building and the surrounding environment will be affected by the choice of construction. When buildings are heated to counteract the cold weather, that heat will gradually be conducted to the outside through the building fabric, as well as in the warm air that leaks through gaps or flows through doors or windows. Since this heat must be replaced in order to maintain a comfortable internal temperature, the rate at which heat is lost will determine the amount of energy used by the heating appliance to maintain a steady temperature, which will in turn determine how much it will cost to run. Similarly, when buildings are cooled using air-conditioning, heat from the outside will gradually find its way in, requiring more heat to be removed and adding to energy use and running costs.

U values

The effect of different construction types upon the rate of heat loss or gain from inside a building can be estimated using a simple calculation. With walls, floors and roofs generally made up of layers of various materials, each layer will contribute to resisting the flow of heat from one side to the other. Each layer will have a particular *resistance*, or 'R' value, where R is equal to the general tendency of the material to resist heat (resistivity, or 'r') multiplied by the thickness of the layer (R = r × t). (Some materials, especially those used for providing insulation, will have R values clearly specified by the manufacturer.) The overall resistance will also be affected by any air gaps or cavities between layers, as well as by surface effects caused by a thin layer of air that adheres to any surface. Since resistance is cumulative, the resistance of each layer can be added together, including the resistance of air gaps and surface layers, to give an overall or total resistance (R_{total}) for a particular wall, floor or roof. While it is helpful to know the total resistance, a more useful figure when calculating the rate of heat loss is the *total conductance*, or 'U' value (K_{total}) of a given construction type.[12] Conductances are not cumulative, and thus cannot be added, so that the overall conductance must be found by inverting the total resistance ($K_{total} = 1/R_{total}$). Although U values are difficult to calculate, a range of examples can be found in the *New Metric Handbook*[13] that can be used to approximate similar construction types.

Design heat loss/gain rate

U values can be used to compare the thermal performance of different construction types and can also be used to estimate the overall energy use for a particular building, as shown in Appendix 1 (page 101). The surface area of each element (wall, floor, roof, window) can be multiplied by its U value, with the results added together to obtain an overall figure known as the *design heat loss/gain rate*. The figure can also take into account heat loss or gain due to ventilation, based on an estimate of the number of air changes per hour. The figure will not be entirely accurate, since it does not take into account detailed effects such as thermal bridging or microclimatic factors specific to a particular site. However, it does allow comparative measures between different designs, taking account of both the size of any building in terms of its surface area, as well as the proportion of different materials, especially the amount of glazing relative to overall wall area.

Glazing and insulation

The highest U values tend to occur with single-layer glass, indicating that windows present a 'weak point' that provides minimal resistance to the flow of heat. Double glazing improves the resistance by capturing a layer of air between two panes of glass that is too narrow to transfer heat by convection. The lowest U values usually occur with construction types that include layers of material of high resistance, or R value. Bulk insulation is the most common form of high-resistance material, which, because of its light weight, can be easily added between other layers. Insulation works by holding pockets of still air in place, like the down of young birds that humans collect for making lightweight jackets and bed covers. When air is able to move, it can quickly transfer heat through convection, but when held in place, its high resistance helps to trap heat inside. If insulation gets wet, however, it will no longer be effective, since the air gaps can be easily filled with water, which has a much lower resistivity than air.

Compact heat zones

Calculations of energy use based upon the design heat loss/gain rate of a dwelling or other building assume that heat is contained by the external building fabric of ground floor, walls and roof. This means that it is easier to heat smaller, more compact buildings, especially when window area is kept to a minimum, or when attached to adjacent buildings using party walls. This is because there is less external surface through which heat can be lost, and a smaller internal volume of air that needs to be heated. Even if only one room is heated, heat tends to disperse to other rooms, since internal walls are rarely insulated and provide little resistance to the flow of heat. This typically occurs when heat spreads to bedrooms adjacent to or above living areas, which can be beneficial by preventing the need to heat these rooms separately. To reduce heating costs, it is possible to designate 'heating zones' within a building by insulating internal partitions and allowing spaces to be separated using well-sealed doors.

Thermal mass

As well as resisting the flow of heat from one side of a wall to another, some building elements have the capacity to store or retain large amounts of heat. The term *thermal mass* is usually applied to a building element that is both large in size and is made of material with a high specific heat capacity. A large masonry fireplace, a thick concrete slab or a heavy rammed-earth wall can act to absorb or store large amounts of heat within a building. Thermal mass exhibits thermal *inertia*, taking a long time to heat up and cool down, and taking in and releasing large amounts of heat without large changes in temperature. Thermal mass can be used to absorb heat from the inside, stabilizing the internal temperature or storing heat for later use. A slab exposed to the sun can still feel warm several hours later; a fireplace can absorb heat from a fire and gradually re-radiate it into the room. Thermal mass can also be used to absorb heat from the outside, delaying the impact of high temperatures. On a hot summer's day, the interior of a stone building can still feel as cool as the night before, as the walls soak up heat that would otherwise pass to the building interior.

Composite thermal performance

In understanding thermal performance, it is important to note that the ability of a material to *contain* heat, measured by its specific heat capacity, is different from its ability to *conduct* heat, measured by the conductivity. Water, for example, has a high specific heat capacity but low conductivity (making it suitable for using as thermal mass), while metals have a high specific heat capacity and a high conductivity. Many buildings today are made using lightweight construction, which has different thermal properties compared to heavyweight, load-bearing masonry. Well-insulated, lightweight walls will resist the flow of heat from one side to the other, but will not retain large amounts of heat. Heavyweight walls will also resist heat flow, but in addition can absorb heat and then re-radiate it later, which can *delay* the flow of heat from one side to the other. It is possible for two walls, one masonry and one of lightweight construction, to have the same resistance value, but different heat capacity. In climates where temperatures fluctuate throughout the day, heavy walls can 'dampen' or reduce the extremes by delaying their impact on interior space. But in situations where temperatures are reasonably constant, including buildings that are heated or cooled artificially, lightweight walls can help to maintain a temperature difference without absorbing heat from either inside or out. In temperate climates, composite construction can bring together the benefits of thermal mass and insulation. Using *reverse brick veneer*, an internal layer of masonry can combine with heating and cooling systems to stabilize internal temperatures, while an outer layer of lightweight insulation can protect the masonry from extremes of temperature variation. However, the strength and durability of brickwork mean that it is generally considered as a material for external use, and insulation is more easily protected behind internal linings. Thus standard brick veneer, although less thermally efficient, is one of the most common construction types for domestic buildings.

chapter three

LIGHT

I installed a skylight in my apartment ...
The people who live above me are furious!

Steven Wright

It is perhaps not too radical to suggest that without light there would be no architecture. Architecture is drawn from light; light helps us to experience architecture as form and space, material and colour, structure and detail. Light *reveals* architecture as both meaningful form and inhabitable space.[1] Light helps us to navigate our way in and around buildings, and it helps us to perform the various activities for which buildings are made. Light is at once poetic and pragmatic and, for many architects, is the most significant ingredient in the making of space. Light is not simply an effect that happens to buildings once they are built, but is a fundamental element out of which architecture is designed and constructed.[2]

Forms of light

The principal source of light is the sun, which provides natural light directly, as sunlight, and indirectly, as daylight. Sunlight and daylight together illuminate building interiors by passing through windows. When light is not available from the sun, it can be created by combustion, by burning wood or wax, oil or gas. Or it can be created through the artifice of electricity, using lamps to convert energy into light. Together, natural and artificial light give variety and flexibility to the way buildings are seen, the way they are used and the way they are designed.

Natural light

The sun is a moving source, providing direct light and heat from its ever-changing position in the sky as it cycles through the days and seasons. When light from the sun reaches the atmosphere, it scatters to become the blue of the sky and the white or grey of clouds. The combination of direct sunlight and indirect daylight creates complex patterns of light that are an essential part of daily life. Even though sunlight and daylight constantly change, they are usually experienced as light from above, illuminating objects around us on the surface of the Earth. Human eyes have a particular sensitivity to the spectral frequency of sunlight, and the human face, with deep-set eye sockets and protecting brow, has evolved to shield the eyes from overhead light and focus attention towards the horizon.

Light and openings

Buildings, too, tend to have a horizontal focus in relation to light. Since most buildings are designed with a roof to provide protection from rain and from the heat of overhead sun, both sunlight and daylight tend to enter laterally through windows or other openings in walls. Windows also respond to the horizontality of vision, enabling views of a nearby garden or landscape, or of an urban environment brought to life by the activities of people. Despite this lateral focus, most windows will allow some part of the sky to be seen from the building interior, which in turn acts as a light source for the space. Fortunately, the sky is relatively bright, and our eyes quite sensitive, so that even the small amount that can be seen through a single window will often be enough to light a room. Natural light will be reduced if the sky visible through a window is blocked by trees or other buildings, or increased if the window admits light from the sun. Despite the variation in the position of the sun and the amount of cloud, daylight provides a reasonably even light source, with the dome of the sky providing diffuse light from every direction.

Light and structure

Another reason for locating openings in walls is that they are easier to construct, and easier to protect against the weather, than openings in a roof. In a load-bearing wall, the basic form of window is a narrow opening, limited by the span of the lintel or arch at the top, extending from above head height down to a sill at the bottom. Such a window is usually located away from corners, towards the centre of a space, in order to minimize the impact on the structural integrity of the wall. In larger spaces, when more light is needed, multiple openings are generally used, often alternating with the structural rhythm of columns carrying loads from above. When the glass is fitted with a frame and hinge, it also allows the window to be opened or closed to take advantage of the weather outside. Load-bearing walls can also be tapered to create a reveal, enabling the light to be reflected on its way into the space, creating a partially lit surface between the bright windowpane and the darker wall that can help to reduce glare.

Source and surface

Windows and light fittings are sources of light that can be used to illuminate interior space. From these sources, light will distribute through that space by reflecting from various objects and surfaces, depending on their colour and texture, and depending on their position in relation to the sources of light. The direct light from sources helps us to see the shape of objects by lighting them unevenly; the indirect light caused by reflection helps us to see objects in their entirety. Light sources vary in size and shape and in the colour and intensity of the light they emit. Surfaces may be translucent, transparent or opaque, and will reflect or redirect light depending on their shape and pattern and on the colour and texture of their materials and finishes. Highly polished or reflective surfaces, such as water, marble, mirrors and glass, will also create *specular* reflections, multiplying the image of objects and light sources as seen within a space.[3] The visual complexity and variety of built space comes from an almost infinite combination of source and surface, natural and artificial, direct and indirect light.

Light and glass

In the past several centuries, technologies for both natural and artificial light have changed dramatically, facilitated in particular by the remarkable properties of glass.[4] Changes in materials and finishes available for construction have affected the behaviour of light, by changing the way it is reflected, absorbed or transmitted by the surfaces that define built space. Advances in structure, framing and waterproofing have changed the design of windows, while the increasing availability of glass has affected the size of openings that can be created. Glass makes buildings transparent, bringing light and views while giving protection from dust, insects and noise. But as a building material, glass has very little tensile or compressive strength, cannot easily be worked on-site, and is difficult to cut or fix to other materials. Although available now in large sheets, glass was originally made by hand in small discs or 'lights' that needed to be joined together, held in place by strips of metal such as lead (hence *leadlight*). Because it is brittle, glass is usually held in a frame, sealed around the edges by more workable materials such as timber or metal. When held vertically, glass will be partially cleaned by rain flowing down the surface, picking up dirt as it goes (sometimes known as 'self-cleaning'). Using glass to seal openings in a roof is more problematic. Glass at an angle is more likely to be broken by falling objects and cannot support the weight of a person during access for maintenance. It is harder to waterproof, since the water must be channelled through framing to reach gutters and downpipes. It also less likely to be kept clean by rain, which flows around dirt as it

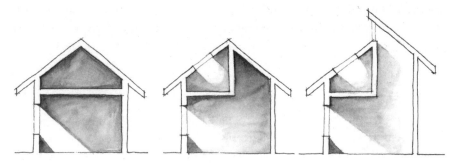

Figure 3.1 Window, skylight and clerestory lighting

forms into rivulets across the surface. These problems have traditionally been avoided by constructing 'saw-tooth' or 'lantern' type roof openings where the glass is held vertically.

Despite the difficulties of using overhead glass, it has become a common architectural device, the light from above complementing the lateral glass of windows. Light from the sky overhead is several times brighter than at the horizon and is less likely to be blocked by trees or other buildings. Glass facing upwards is more likely to admit sunlight, which can result in problems with heat gain, but can also create interesting effects as patches of sunlight reflect from high-level surfaces to bring even more light into a space. Fully glazed roofs can also provide an intermediate space between the enclosure of an interior and the exposure of external space. From shopping arcades to the atria now popular in office buildings, the glass roof creates a controlled environment protected from the wind and rain and isolated from extremes of heat or cold.

Floorplate depth

Giving each space in a building access to an external wall for daylighting and ventilation is easy at domestic scale, but for larger buildings, the depth of the floorplate can affect the available light. The amount of daylight entering from a window gradually diminishes towards the building interior, as the relative view of the sky decreases. However, the problem is not simply the distance from the window, but the depth of the space relative to the height of the window. In general, daylight levels in a space are reasonably good for a distance several times the window height back into the room; around 6–8 metres for a typical window.[5] But it is the proportion of the room and its fenestration, and not the actual dimensions, that determine the quality of the light. Larger floorplates simply need higher ceilings, allowing taller windows to bring light into the space. Clerestory windows, located a 'clear-storey' above ground level, allow light to be brought deep into a building interior. In religious buildings, this high-level lighting is often exaggerated, and ground-level windows omitted, giving a vertical emphasis to space and expressing the divine quality of light as it reaches down to Earth.[6] In multi-storey buildings, the distance from a window is much more of a problem, since each floor is covered by the floor above, with floor-to-ceiling heights typically less than 3 metres. Some European countries recommend a maximum distance that office workers should be located away from windows of about 6 metres. This involves limiting floorplate dimensions, or bringing light down through atria or void spaces, to provide access to natural light.

Light and movement

The quality of daylight in any space is determined primarily by the amount of external building fabric available for glazing, whether from the side or from above. If light is available from more than one direction, it will multiply the effects of changing

Figure 3.2 Light above a staircase to encourage upward movement

light, emphasizing the sense of time that comes from connection with the outdoor environment.[7] And since functional relationships of spaces in plan may limit the external access for any given space, one of the key determinants of plan form, massing and building envelope is the need to ensure good daylighting in each space. Light is also essential to planning layouts because of its role in circulation. Although the amount of light needed for circulation is much less than for other tasks, light has a significant effect on the way people move through a building. In general, people are *phototropic*; that is, they are attracted towards light. Once inside a building, people tend to move towards the building edge, where windows let in daylight and sunlight and give views to the exterior. Directing people to the centre of a building can also be done with light. Using an atrium or a glazed roof, especially over a staircase where openings are required in each floorplate, light can be brought down into a building interior to guide people in and up to spaces above (see Figure 3.2).

Steel and glass

The patterns of fenestration and circulation in buildings, initially constrained by load-bearing construction, have been transformed by the use of glass and steel since the late 19th century. Steel can be used to make rigid frames, which frees the wall from carrying structural loads, reducing its role in weather protection. Combined with float glass technology, this meant that entire walls can be made of glass, suspended – like a curtain – from above. Supported by artificial systems for ventilation and temperature control, the external skin, as Reyner Banham observed, does little more than prevent people from falling out.[8] The glass curtain wall creates a strange hybrid, the *window-wall*, with neither the flexibility of the window nor the solidity of the wall, and yet providing both openness and enclosure. The familiar pattern of the facade as a composition of solid and void is instead replaced by a sheer, flat surface broken only by frames or joint lines between individual sheets of glass. The recognition of scale and depth that normally results from window patterns and facade details can be lost when buildings are fully glazed, which can make them appear monolithic and alienating.[9] Although valued for its transparency, the use of tinted or reflective coatings on glass to reduce heat load can make buildings appear opaque and unwelcoming, a private or secretive space cut off from engagement with urban space. However, recent developments in glass colours and coatings, including *phase change* glass, which can convert between transparent and opaque by the addition of an electric current, give rise to immense opportunities for filtering light. Glass can also be protected by finely perforated sunshading screens, creating layered and blurred boundaries that appear on the inside as complex shadows during the day, and on the outside, like a glowing lantern at night.

Measuring daylight

The easiest way to anticipate the effects of daylight within a building while it is being designed is to construct models, since apart from changes in texture or material, the scale does not alter the physical behaviour of the light. As well as building models, there are various calculation methods available for estimating the amount of daylight that will be available within a building. These involve considering the dome of the sky as light source, which varies throughout the day and year depending on the position of the sun and on the amount of cloud cover. Because of this variation, estimates usually assume a uniformly overcast sky, the most commonly used being the CIE (Commission Internationale de l'Eclairage) standard sky, which is defined as being three times brighter overhead than at the horizon. Daylight figures are available that describe the *illuminance*, or amount of light falling on a surface such as a table or the page of a book, from an unobstructed sky. Daylight figures for London, for example, are usually assumed to be at least 5000 lux for 90 per cent of the time between normal working hours of 9 am to 5 pm.[10]

At any point inside a building, however, much of the light from the sky will be blocked. Since people can read comfortably at an illuminance level of around 300–400 lux, only a small percentage of the sky, around 5–10 per cent, needs to be visible from any point in a building to enable reading to occur. Calculation methods involve determining the amount of sky that can still be seen from any given point inside a building. Before the use of computers in design, this was complicated by the need to convert between the orthogonal geometry of building plans and sections and the spherical geometry of the sky. Throughout the 20th century, a range of manual techniques for estimating lighting levels were developed, using charts, tables, protractors and methods of projection to overcome this problem. Methods including the CIE Chart, the BRS Protractor, the Pleijel Diagram and the Waldram Diagram enable daylight levels to be calculated using only plan and section drawings.[11] Each method involves a way of counting up how much of the sky would be visible from a given point in a room based upon window geometry. Today, software programs such as Radiance make the task easier and are able to calculate the lighting level at each point in a room and display them using *isolux*, or equal light, contour lines.

Such calculations are helpful, but are rarely undertaken at design stage. Most architects rely on an intuitive judgement regarding the amount of window needed in relation to the size and shape of any space, and windows that are too small for daylight are often too small for other purposes, especially view. And since artificial lighting is usually installed to make spaces usable at night, minimum lighting requirements tend to be satisfied by artificial rather than natural light. And while many building codes specify the need for any habitable room to be equipped with a window, there is increasing tolerance for the use of borrowed light, where rooms with no external walls gain light via glazed partitions from adjacent rooms.

The ease of using glass means that the problem of having too little daylight may be less likely to arise than the problem of having too much. The dome of the sky can be a very bright light source and, with poorly designed windows, can lead to glare. Glare is a form of visual discomfort caused by excessive contrast, where brightness levels in the field of view are too far apart for the eye to adapt easily. Ideally, the source of light should be outside the field of view, above or behind the eye, lighting objects in front of it. But when looking towards a single light source such as a large window, the brightness of the sky can overwhelm the objects silhouetted in front of it. This is sometimes used intentionally in religious architecture, lighting the altar from behind in a glow of light. But if unwanted, it can be avoided by filtering the light or by dispersing it more evenly throughout the space. This can be done by creating reveals or locating windows against a side wall, or by using light shelves to bounce light from a window up on to the ceiling and back into the space. Or it can be done using multiple sources of light, natural as well as artificial, throughout any space.

Artificial light

Artificial light can never replace the singular power of the sun as the source of light for daily life. But what it lacks in authenticity, it more than makes up for in versatility. Artificial lights are like tiny suns: able to be placed anywhere, turned on and off at will, and available in an infinite range of colour and brightness. With a portable light source such as a desk lamp, it is possible to bring light to a space by simply plugging in to a power outlet. But light sources work better when placed high in a space to mimic the overhead quality of natural light. This is a problem if the light comes from fire, with chandeliers, for example, originally designed to be raised and lowered so that the candles could be lit and extinguished. With electric lighting, however, a wire can be brought down within reach and connected to a switch. But because the wires contain enough current to kill a person, light fittings must be installed by a licensed electrician and cannot simply be added by the occupants to adjust the lighting.

The need to install light fittings provides an opportunity to integrate them with the spatial and formal qualities of any building. It is easy to add a grid of lights to achieve sufficient lighting levels, but far better to use the lights to define particular spaces, highlight distinctive surfaces or emphasize the rhythm of the structure. Lighting can be integrated with architecture, concealed in niches or soffits, or exposed in fittings to complement and create patterns of form, space and light. It is one thing to place lights in a space; it is another to use light as an integral part of spatial organization.

Light from fire

The first form of 'artificial' light was, of course, fire. By producing light as well as heat, fire transformed buildings from mere shelter into places for human culture and habitation. The light from fire helps to overcome darkness, its warm, low light creating a focus (literally a *hearth*) for myth, ritual and dreams. 'If fire,' writes Gaston Bachelard, 'was taken to be a constituent element of the Universe, is it not because it is an element of human thought, the prime element of reverie?'[12] Just as the heat from fire can be tamed, so too can its light. Candles, oil lamps and gas lamps provided a controlled means of indoor burning of carbon-based fuels. These brought associated dangers of setting buildings alight and also caused combustion by-products to be deposited on internal surfaces. A 'spring clean' was originally necessary to remove the deposits caused by coal fires and oil lamps used in winter. Dark and heavy furnishings, such as those of Victorian England, were used to disguise the build-up of soot. And because candles and lamps needed to be lit and extinguished, they were mostly positioned within reach on tables or low wall fittings, reducing the risk of setting fire to the ceiling.

Incandescent light

In the 19th century, the development of electricity enabled combustion to be separated from application of the energy released. Coal-fired power stations could be used to generate electricity, which could then be converted into heat or light inside houses without the dirt or danger of fire. But even the first light bulbs worked on a principle similar to that of fire: the 'glowing' of a metal filament heated up by electric current, known as *incandescence*. Thomas Edison's patent for the electric light bulb was part of a strategy of encouraging customers to connect to his Edison General Electric Company, which eventually became General Electric, one of the largest companies in the world. Electric lighting soon became used in a variety of situations, from street lights, to motor vehicles, to advertising signs, as well as on the inside and outside of virtually every building. The high cost of installation meant that electricity was initially used only for street lighting or for commercial uses such as theatres or shop windows.[13] Another major use was for advertising, enabling billboards to be lit at night and even incorporating light bulbs into the signs. The later development of neon light further enhanced the impact of advertising by adding a range of vivid colours. The largest installations, such as New York's Times Square, soon became attractions for visitors, with cities such as Las Vegas and Tokyo eventually becoming famous for the spectacle of night lighting.

The light from incandescent lamps is similar to that emitted by fire; the warm, yellowish light containing frequencies across the visible spectrum, which makes skin look healthy but creates a lot of heat. Like fire, incandescent bulbs emit mostly heat; only a few per cent of the energy used is converted into light. This originally made large-scale installations unfeasible, especially for indoor use. Because of their low cost and warm light, incandescent lamps continue to be used in domestic situations, even though lamp life is relatively short, lasting only about 1000 hours before needing replacement. More advanced forms of incandescent bulbs are now available, in particular low-voltage tungsten-halogen (or quartz-halogen) lamps. These produce about twice as much light for the same amount of energy as standard incandescent lamps, and last about twice as long. Halogen lamps are much smaller than standard incandescent lamps, making them more versatile for detailed lighting designs such as for task lighting or retail display.

Fluorescent lighting

With incandescent lamps, the absence of flame and remote switching meant that they could be installed at ceiling level. But it was only with the development of the fluorescent lamp that artificial lighting was used to substitute for daylight. Fluorescent lighting involves passing an electric current through mercury vapour inside a glass tube, causing it to emit ultraviolet light; which in turn causes the coating on the tube, known as a *phosphor*, to emit visible light. This is emitted as 'spikes' of light at particular wavelengths, rather than as a broad spectrum of light and heat. Early fluorescent lamps, in particular, produced mostly bluish light, with almost no yellow to make the light appear warm and pick up essential skin tones. More recent versions have a better mix of wavelengths and produce light similar to daylight.

Fluorescent lamps are far more efficient, producing about five times as much light and lasting about five times longer, than standard incandescent lamps. They also create much less heat and can be easily installed at ceiling level. This greater efficiency, instead of leading to reduced energy use, led to more extensive installations, creating 'artificial' daylight conditions, but without the need for large windows or high ceilings (see Figure 3.3). Fluorescent lighting emerged in parallel with other technologies, especially air-conditioning, that liberated indoor space from its reliance on the external environment. It is these technologies that made high-rise buildings with low ceilings and deep floorplates the norm for a growing number of office workers performing paper-based tasks rather than manual labour. Artificial light was promoted as being more reliable than daylight and able to make workers more 'productive' and efficient. Interestingly, manufacturers of fluorescent lighting were often involved in determining the light levels that were recommended in national standards.[14]

Glass curtain wall towers transformed the appearance of cities, appearing opaque by day, but inverting at night as rows of fluorescent tubes became visible from the outside, lighting up the buildings like a lantern. Today, we are used to the cinematic spectacle of the city, combining the lights of streets and traffic with advertising and buildings. But the use of electric light is so pervasive that cities are often devoid of darkness, bathed in a glow of light that makes it difficult to see the stars in the night sky. Various protest movements have arisen, calling for a reduction of light pollution, not just to save energy but also to recover the experience of darkness.

Today, the standard light fitting for indoor use in commercial and institutional situations remains the fluorescent tube. Even when daylight is available, artificial

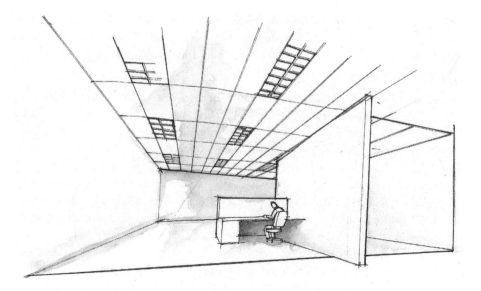

Figure 3.3 Artificial light as daylight

lights are installed to improve lighting quality and to ensure that spaces can be used at night. It is also possible to have rooms without any windows at all, or in dense urban situations, where windows are blocked by adjacent buildings. Spaces that are occupied for short periods of time, such as toilets, or where full lighting control is needed, such as lecture theatres, can be located within a deep floorplate having no access to an external wall. This makes for larger building dimensions and more versatile planning arrangements, but it can also limit the sense of outside connection for building occupants.

Measuring artificial light

The design of artificial lighting can be used to mimic either the dramatic intensity of firelight or the even distribution of daylight. It can involve the use of decorative light fittings as a form of ornament within a space, or the concealing of lamps to highlight particular surfaces or features, which in turn act as a light source as they redirect light into the space. Or a light source can be suspended from, or recessed within, a ceiling to create an even distribution of light for work surfaces below. Each of these different uses involves a choice of *luminaire*, or light fitting, in which the lamp (light bulb) is held. An immense variety of luminaires is available, including pendants (hanging from a ceiling), wall lights, recessed downlights, uplights, spotlights, task lights and concealed lights. Installing bare lamps (light bulbs) without luminaires will usually cause glare, as the source of light may fall within the field of view of people in the space, making it harder for them to see the objects that are the true focus of their attention. Luminaires are needed primarily to prevent glare by diffusing or shielding the light source from view, as well as for directing and distributing the light to where it is needed throughout the space.

Luminaires provide a fitting to hold the lamp (or bulb) that converts electrical energy into light. The amount of light emitted by a particular lamp in a given amount of time is known as the *luminous flux* and is measured in *lumens*. The amount of light reaching a work surface below is known as the *illuminance*, which is measured in *lux*. Illuminance is equivalent to light flow (luminous flux) per unit area, with one lux equal to one lumen per square metre. The amount of light required in a space depends upon the tasks likely to be undertaken. Rooms for circulation or storage only need 50–100 lux; rooms for reading in schools, homes or offices need around 240–400 lux; and areas for detailed or intricate tasks such as jewellery making may need up to 1600 lux.[15] Required lighting levels can be used to determine how many luminaires need to be installed, using simple calculation methods. Most lighting suppliers or manufacturers will do this when you buy their products, or a lighting consultant can be used. Since each lamp will vary in the amount of light it gives out, and since each luminaire will vary in how it distributes that light, calculations depend upon relevant manufacturer's data being available. Data for each lamp should include a value for luminous flux.[16] A typical incandescent bulb will emit around 1000–2000 lumens; a typical fluorescent tube will emit around 3000–6000 lumens.

Calculating how much light must enter a space to achieve a suitable lighting level, and thus how many lamps or bulbs must be installed, is simply a matter of multiplying the required illuminance (in lux) by the area of the space. However, not all of the light from the lamps will make its way down to the work surface. There will be some loss of light due to maintenance, depending on how much dirt is in the space and depending on how often the lamps are cleaned and replaced when faulty. There will also be some loss of light to the surfaces of the room, depending on whether they are light or dark and depending on the geometry of the room, which affects how much wall surface the light may encounter on its way down from the luminaire to the work surface. The number of lamps required must be factored up to account for these losses. The first, known as the *maintenance factor* (MF), can often be ignored for office spaces with regular cleaning schedules. But the second, known as the *utilization factor* (UF), depends on the design of the luminaire and must be derived from manufacturer's tables, taking account of finishes and proportion of the space. This is known as the *lumen method*, which is described in Appendix 2 (page 103).

Many lighting installations, especially in office space, are meant to encourage flexibility by ensuring an even lighting at every point in the space, usually at a level suitable for paper-based activity. Unfortunately, this can lead to a fairly monotonous and harsh lighting environment, devoid of the subtle variation that occurs with natural lighting. Recently, the idea that fluorescent light can entirely replace daylight in all situations, especially extended hours of office work, is no longer valid, and 'access' to daylight for workers is improving.[17] Also, the need for a variety of different spaces for different tasks has led to more varied lighting, generally with a lower background level and brighter task lighting only where it is needed. This is not only more appropriate for computer-based work in which the screen is itself the light source, it also provides greater lighting 'texture' allowing the eyes a chance to adjust regularly and avoid fatigue. This sort of irregular lighting design is not only better for vision, but can provide design opportunities by avoiding the need for regular grids and allowing more complex patterns of lighting levels, fixtures and finishes.

These changes, intended to create a more varied visual environment, are also contributing to improved energy efficiency of lighting installations. Lower lighting levels, combined with more efficient electronic ballasts and smaller diameter T5 fluorescent lamps, reduce the amount of energy used for artificial lighting. Recent changes to building codes will make such efficiencies mandatory. New energy provisions specify maximum values for the rate of energy use per unit area, known as the *illumination power density*, for various spaces depending on use. This will shift the focus away from minimum lighting levels and instead encourage greater integration between natural and artificial light.

Colour

Light is made up of colour, and the play of light in and around buildings is enhanced and multiplied as that light is scattered into its various colours. But colour has a complicated role in architecture, its use affected by available materials, pigments and light sources, as well as by changing attitudes to decoration or ornament. Because colour is often applied to building elements as part of a surface layer or protective coating, it is often seen to have no inherent function or purpose, only to be changed many times throughout the life of a building in accordance with shifting fashion or taste. Colour may also be considered as part of the decoration or ornament applied to the surfaces of a building and not an inherent part of the building fabric. But colour does have a significant role to play in the function, appearance, symbolism and mood of built space.

Dimensions of colour

In 1665, Isaac Newton proved that light was made of colour. By using two prisms, he separated white light into the colours of the spectrum and then merged them back again into white light. The colour of light depends on the combination of wavelengths emitted from a light source or reflected from a surface. A light source will appear white only if all the wavelengths are present; a source that emits only a single wavelength, or a limited combination, will appear coloured. A surface will absorb some wavelengths of light and reflect others, so that the colour of the surface depends on the particular wavelengths that are reflected. The colour from a light source is said to be *additive*, while the colour from surfaces is said to be *subtractive*. A surface can only reflect light that falls upon it, so that while a surface will have a characteristic colour under white light, it can be made to appear different when viewed under coloured light. Good colour rendering, enabling the characteristic colour of a surface to be seen accurately, will only occur if the light falling on it has an even combination of wavelengths making up white light.

Inherent and applied colour

In many works of architecture, the use of colour is avoided, or at least played down by using dull or muted colours. Colour is often left to building interiors, applied as paints or decorations that are less susceptible to fading by the sun and that can be changed more readily to suit the requirements of the occupants. For external surfaces, architects will often rely on the 'natural' colours of materials such as stone, marble or timber. However, many materials are chosen precisely for their inherent colours, such as gold leaf or marble, or applied in patterns, such as polychromic brickwork or terrazzo floors. Other materials, such as concrete or glass, can readily be altered by adding ingredients or pigments of a particular colour, depending on available technology. Materials such as steel and timber usually require coatings to protect them from the weather, which will often perform better when colours are added. The avoidance of colour was a defining characteristic of the Modern movement, with white walls and glass providing clean lines and bright internal spaces. Fortunately, Modernist architecture was not entirely colourless, with experiments in primary colours of red, yellow and blue characterizing the work of the de Stijl movement in the Netherlands, especially Schröder House in Utrecht by Gerrit Rietveld (1925). Colour was also used to modify light, such as Le Corbusier's use of stained-glass windows in the Chapel at Ronchamp (1957) and in the convent at La Tourette (1960). Other architects extended the limited palette of Modernism by adapting it to suit local traditions, such as with Luis Barragan's villas in Mexico, featuring walls rendered in bright colours of pink, yellow, blue and purple.

Colour coding

Colour's significance for architecture goes beyond its role as ornament or decoration. The choice of colour can make a space feel warm or cool, light or dark. It can also play a functional role, especially when used for coding of elements or spaces. One of the first major projects of the High-Tech movement, the Centre Pompidou in Paris by Renzo Piano and Richard Rogers (1977), used colour coding for the exposed ductwork, with blue for air, green for water, yellow for electricity and red for escalators and elevators. A similar coding was also used a decade earlier by Le Corbusier in La Maison de l'Homme (also known as the Heidi Weber Pavilion) in Zurich (1967), with red for hydronic heating, yellow for electricity and blue for water. Similar colours were used by James Stirling in the design for the Staatsgalerie in Stuttgart (1984), although without being coded to represent services. Other architects of the postmodern era, such as Michael Graves and Charles Moore in the United States, used vivid combinations of colour and neon light in order to overcome the restraint of Modernism.

Specifying colour

The need to specify colours in order to ensure accurate matching and reproduction is essential to industries such as printing and fashion, as well as for the diverse range of materials and finishes used in architecture. Colours can be specified electronically, with many computer programs today enabling colours to be defined by coordinates on a colour wheel or palette. For many years, this was done using a chromaticity chart developed by the Commission Internationale de l'Eclairage (CIE), in which coordinates specify a position within a field bounded by the three primary colours of red, yellow and blue. Another way of specifying colour is using Munsell colour charts, which define colours according to the three dimensions of hue, saturation and brightness. Hue defines the colour within a spectral range of red, orange, yellow, green, blue, indigo and violet. Saturation defines the amount of colour present, so that, for example, pink and red may be the same hue but different saturation. Brightness defines the amount of black or white pigment, making the colour appear dark or bright.

Colour contrast

The reason for understanding these three dimensions is that they give rise to different forms of colour contrast. The perception of colour is often understood as being subjective, a point emphasized by the small percentage of people who demonstrate colour blindness, unable to distinguish colours that others see as separate, such as red and green or blue and yellow. But as with other senses, the experience of colour is not just different for each person, but can depend on the context; that is, on the relation between colours that are viewed together. The way a particular colour appears is affected by the colours around it, especially when there is contrast in one or more of the dimensions of hue, saturation and brightness. This contrast can be used to emphasize colours, making them seem brighter or more vivid than when viewed alone. Colour mixing usually involves choosing colours that will complement each other by providing appropriate contrast. With hue or colour contrast, colours at opposite sides of the colour wheel can be seen to enhance each other: red and green, purple and yellow, blue and orange. With saturation contrast, intense colours and pale colours can be used together, such as pink and red. With brightness contrast, a dark version and a light version of the same colour can be used, such as pink and brown. Or alternatively, multiple contrast can be used, such as bright red against pale green.

These different forms of contrast appear regularly in the graphic arts, especially print media and fashion, and are now becoming more widely used in architecture. The use of colour in architecture is also facilitated by the diverse range of printed and patterned surfaces, coloured glass and lighting now available. Striking compositions of colour and light can be found in works by Will Alsop in England, Stephen Holl in the United States, Ben van Berkel in the Netherlands, Sauerbruch Hutton in Germany, and Lyons Architects in Australia, to name just a few. Many of these use grids or patterns that appear like sample boards of complementary and contrasting colours. By enriching sources as well as surfaces, colour adds an extra dimension to the quality of light in space. The play of colour from materials and applied finishes, from coloured lights and tinted glass, can enliven the experience of space and enhance the character and meaning of built fabric.

chapter four

SOUND

I didn't get a toy train like the other kids. I got a toy subway instead.
You couldn't see anything, but every now and then you'd hear this rumbling noise go by.

Steven Wright

We are constantly reminded of the dominance of visual culture, with media such as newspapers, magazines, television, cinema and the Internet providing an ever-changing and often unavoidable array of images in front of our eyes. But despite this dominance, aural culture, especially through language and music, continues to be an essential form of communication between people. Technologies of visual culture reconfigure the way words are conveyed from one person to another.[1] New technologies have also transformed sound, as words and music are recorded and transmitted across airwaves and telephone lines. While various artefacts are available for making or controlling sound, there are also a large number of familiar sounds in urban environments that are produced as a by-product of other processes, such as the closing of a door, the flush of a toilet or the 'hum' of a refrigerator.[2]

Sound and space

Buildings have significant consequences for sound and its spatial distribution, as walls can enhance and direct sound within space or protect and isolate space from sounds occurring outside. Buildings contain a diverse range of human activities and machines, each with their unique sounds. Buildings also have their own characteristic sound, as they respond to the movement of people, or wind, or rain. By keeping sound out, walls help to define zones of privacy and quiet activity, from lecture theatres to libraries, from boardrooms to bedrooms. By keeping sound in, walls help to determine the quality and the distribution of sound, both of which are essential to the experience of space. The way in which walls can reduce sound by limiting its transmission from one space to another, or enhance sound by reflecting it within a space, together make up the broad field of architectural acoustics.

Sound and buildings

Sound tends to 'fill' a space; like light, reflections bounce from every surface, making it part of the overall ambience rather than just attaching to the source. Whether originating from outside or in, sounds compete for attention, although people can choose to focus on the sound that is most important to them at the time. Unlike other forms of energy, especially heat and light, people are often the major source of sound in any space. People activate space by the sounds they introduce, whether from speech, movement or activity.[3] When people enter a space for the first time, they may even feel compelled to make a sound to test the impact that their voice will have, to know whether to speak loudly or softly. People measure a space by the sound of their footsteps, the echo or reverberation quickly indicating the scale and temperament of the space. As Juhani Pallasmaa says, 'Every building or space has its characteristic sound of intimacy or monumentality, invitation or rejection, hospitality or hostility.'[4]

The way a building returns sound to its inhabitants reveals whether it is warm and welcoming or cold and austere. But the way a building transmits sounds between inhabitants can also reveal a great deal about the types of material used and how they have been assembled. The sound of people interacting with parts of a building, from footsteps to closing doors, from turning on taps to flushing toilets, can be conveyed through walls and floors, the amount of sound transmitted depending on the quality of their construction. Sound can find its way through gaps or holes, or through poorly sealed windows or doors. Similarly, the creaks and groans as buildings move in the wind quickly draw attention to the stability of the structure. To check construction, we may even hit the walls like a drum to hear the resonance and feel whether lightweight or robust materials have been used. Different materials have different sounds under the impact of the hand or foot: the crunch of gravel, the drum of timber, the rattle of metal sheet.

People measure a space by the way it accepts sound, but can also determine the activities within by the level of sound already contained. Some spaces are designed for all the inhabitants to hear a single sound source, such as a person speaking or playing an instrument, but many others are designed to accommodate multiple sound sources, such as the conversations of patrons in a restaurant or bar. Here, a level of 'background' noise can give a sense of activity in the space, but also contains the sounds of conversation to the space of each group. Too little sound, and privacy is lost; too much sound, and people may have trouble hearing each other across a table.[5]

Sound and noise

Sound usually originates from a point source, emanating outwards like the surface of a sphere and reducing in intensity as the distance from the source increases. Because the surface area of a sphere is derived from the square of its radius (πr^2), the intensity of sound follows the *Inverse Square Law*, such that it decreases in proportion to the square of the distance from the source. Thus if the distance between the listener and the sound source is doubled, there will only be one-quarter as much energy reaching the listener.

Unfortunately, this means that the sound is loudest at the source. Any sound that is unwanted by the listener is defined as noise, suggesting a combination of sound level (too loud) and content (unwanted). Whether from playing music, or operating machines, or simply a loud conversation, people can unwittingly be a source of noise for those around them. Rarely are people disturbed by their own sound, and people making noise are often unaware of the impact their sound has on others – especially when talking with someone on the other end of a telephone.

Another reason that sound can be unwanted is that, like heat, it is often simply a by-product of other mechanical processes. Machines such as air-conditioners and motor vehicles create sound from combustion, friction and vibration from engines, tyres and moving air. Once again, the person causing the sound is often least aware of its impact on others, enclosed within a protected interior environment while the noise used to create it is propagated outside. Many of the machines that create noise in the urban environment did not exist a century ago. Machines make cities more habitable, but have also made them noisier. Major sources of noise include the motor vehicles used for transporting people and goods, for cleaning streets and removing waste. Another major source is the demolition and construction of roads and buildings. Fortunately, many sources of noise are located away from street level, such as underground trains or roof-top air-conditioners. However, the desire to avoid the noise of the streets in many cases motivates the use of sealed glass curtain walls and air-conditioning systems instead of operable windows, keeping sound out and creating quiet space within.

Noise prevention

The sounds of the city result in large part from the sheer density of urban populations. People crowd together in cities, forgoing the open space of the suburbs to take advantage of opportunities for work and recreation at all hours of the day and night. One of the main roles of architecture is to facilitate that density, making space habitable by reducing the impact of people on one another, creating spaces of acoustic and visual privacy among the sounds and view of others. The best way to stop sound is by using heavy walls, or by the distance afforded by intermediate spaces. But where this is not possible, a louder, introduced sound can be used to provide masking, such as the background noise of a water feature, piped music, or even air-conditioning. A similar logic is behind the use of personal music players, enabling an individual zone of privacy to be created in the public space of the city. Masking can also be achieved inadvertently, so that people listening to a radio or television, for example, are less sensitive to outside sounds and can easily turn up the volume. The most common complaint about unwanted sound is that it interrupts sleep, often because people have turned off their own sound sources and have become more aware of the sounds made by others.

The dimensions of sound

Sound is caused by vibration of the air, which is transferred by the eardrum to the nerves of the inner ear. The experience of sound depends on a number of factors, especially the frequency of vibration, the energy level and the duration. The frequency, measured in hertz (Hz), is experienced as *pitch*, or whether a sound is high or low. The energy level, measured in decibels (dB), determines the apparent loudness of the sound. The duration, measured in seconds, determines how long a sound lasts for and, in particular, how it interacts with other sounds that follow it. Because the human ear is able to detect multiple vibrations, any particular sound is usually made up of several different frequencies that are heard together as a single sound, or even a number of different sounds that are heard together at the same time. This enables the harmonious quality of musical instruments or voices, as multiple frequencies reinforce one another to give depth or complexity to sound.

But as well as these three dimensions, sound can be identified as occurring in a particular position or direction relative to the listener. That position may also change in time, indicating a relative movement between the listener and the source, which is also experienced as a change in loudness as the sound source comes closer or moves further away. In extreme cases, that motion can give rise to the *Doppler effect*, where the speed of the source compresses the sound waves, making the frequency appear higher as the source moves towards the listener and lower as it moves away. The direction of sound is complicated by the acoustics of space. A direct line of sight and sound makes it possible to identify the source as the loudest and first heard version of any sound, but subsequent reflections will distribute a sound throughout a space. This can be beneficial for music, but can be problematic for disruptive or unwanted sound.

The final quality that affects the experience of sound is what might be described as its *information content*. Through language and intonation, sound enables communication between people. But, sound also identifies its source, as we recognize the sound of a bird, a lion or a train. This can be vital for safety, as it enables people to be aware of, and hopefully to avoid, things that threaten them. Or it may lead to annoyance, as we sense the presence of unwanted guests: a barking dog, a mosquito, a lawnmower, or even just music that is suited to different tastes.

The various dimensions of sound – pitch, loudness, duration, location and information content – are all affected by the environment in which they occur. Once produced at a source, sound waves move out in each direction until they decay due to friction, or until they are reflected from or absorbed by objects around them. In general, sound is reflected by hard surfaces and absorbed by soft surfaces. By the time sounds reach the listener, they have usually been affected in some way by the surfaces around them. The careful placement of barriers and absorbent surfaces can be used to control where a sound occurs, keeping it in where it is wanted, or directing it away from where it is not wanted. But this also means that surfaces can be used to affect the quality of sound, making it seem louder, or last longer, than it would in the open air. In the art and science of room acoustics, space is essential for conveying sounds from source to listener, with a little adjustment in between.

Room acoustics

The science of acoustics began largely with the work of Wallace Clement Sabine at Harvard University in the late 19th century. In 1895, Sabine was asked to look into the problems at the newly built Fogg Theatre, where students were having trouble hearing their lectures. Drawing comparisons with the nearby Sanders Theatre, Sabin moved materials, including seat cushions and oriental rugs, and people back and forth between the two theatres, measuring the time taken for the sound of a human voice, at about 60 decibels, to *decay*, or no longer be audible in the space. This decay time is essential for speech audibility, since it determines whether each new syllable can be heard over the previous one. Sabine called this the *reverberation time*, which he defined as the time taken for a sound to decrease by 60 decibels, from audible to below the threshold of hearing. The formula that he devised is $RT = 0.16 \, V/A$, which shows that the reverberation time depends on the volume (V) of the space, as well as on the amount of absorption (A) caused by the surfaces in the space. Although Sabine originally used cushions, he later defined a more accurate measure of absorption as an open window: a unit of absorption, known as the *sabin*, is equivalent to one square foot of open window.

The total absorption (A) must be found by adding together the area of each surface multiplied by its coefficient of absorption, a number between 0 and 1, where 0 is fully reflective and 1 is fully absorptive. Since coefficients may vary for different frequencies, they may be averaged over a range of frequencies (250, 500, 1000 and 2000 hertz) to give a noise reduction coefficient (NRC). Table 4.1 shows noise reduction coefficients for common materials. More reflective surfaces will give a higher reverberation time; more absorptive surfaces will reduce the reverberation time. Absorbent surfaces include fabrics, felts and even the clothing worn by people, since the fibres of the fabric can vibrate to absorb the sound. The use of padded seats in theatres is more likely to be for acoustic purposes than for comfort, especially when low audience numbers reduce the overall absorption.

Table 4.1 Noise reduction coefficients for common construction materials (after Szokolay)

Material	Frequency		
	125Hz	500Hz	2000Hz
Exposed brickwork	0.05	0.02	0.05
Plasterboard walls	0.2	0.1	0.04
Plywood panel walls	0.3	0.15	0.1
Windows (glass)	0.1	0.04	0.02
Curtains	0.05	0.3	0.5
Ceramic tiles	0.01	0.01	0.02
Cork or vinyl floor	0.02	0.05	0.1
Concrete floor	0.03	0.03	0.05
Medium pile carpet	0.05	0.15	0.45
Ceiling tiles	0.23	0.5	0.5
Acoustic panels (perforated plywood over glass wool)	0.15	0.75	0.75

Sabine managed to improve speech intelligibility for the Fogg Theatre by increasing the amount of absorptive surfaces in the space. He also calculated reverberation times for other theatres, concluding that spaces for speech needed a lower reverberation time, at about 1 second, than spaces for music, which needed reverberation times of about 2 seconds. This is because music can tolerate more 'overlap' between consecutive notes, making it seem warmer and more resonant. The original reverberation time in the Fogg Theatre had been more than 5 seconds, which was experienced as excessive overlap or echo. Reverberation times are still one of the major determinants of the acoustic quality of theatres and other spaces, along with consideration of sound distribution due to room geometry.

Theatre design

The design of theatres based on the physics of sound distribution was one of the major strategies for the functional determination of form in Modernism.[6] The shape of a theatre is usually intended to mimic the distribution of sound waves from the mouth of the speaker, radiating outwards in a sphere, but more audible from the front than the rear. This is the basis for the classical use of circular geometry in outdoor theatres. By stepping the theatre in sections, sight lines were improved, as audibility is assisted when people can see the speaker.

Most theatres today still use a circular or partly circular geometry in plan, with people arranged into rows equidistant from the speaker, as far back as the sound remains audible. (Arranging the audience in a spherical pattern would be good, but probably not very comfortable.) By placing reflective surfaces above and behind the speaker, sound that would have been lost to the rear is captured and 'folded' back towards the audience. The effect is similar to standing in front of several mirrors at once, except that where reflected images are seen as separate, reflected sounds merge together to create a single voice that sounds louder and more resonant than the original (see Figure 4.1).

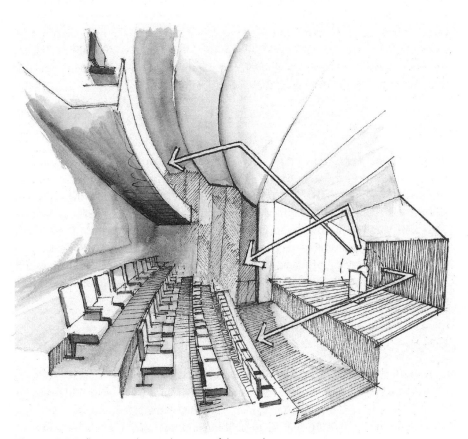

Figure 4.1 Reflections enhance the voice of the speaker

The reflective surfaces make the space act like a megaphone, channelling the voice of the speaker back into the space. Reflections from surfaces a long way from the speaker, especially rear walls, are less likely to provide reinforcement. These are likely to be heard too late after the original, and from the wrong direction, and can thus be perceived as an echo. In spaces intended for music, reflective walls may be faceted into a series of smaller surfaces in order to pick up the high frequency sounds that can easily be lost. A series of small reflective surfaces, such as the highly ornate walls of many older theatres, can create ideal acoustic conditions for music.

Of course, most theatres today use electronic amplification to support the natural patterns of sound distribution, with active and passive acoustics both dealt with by specialist acoustic engineers. But reverberation times and sound distribution can be considered for spaces other than theatres, especially when large numbers of people are likely to be in the same space. In particular, a suitable mix of reflective and absorptive surfaces is essential for determining the acoustic quality of spaces such as offices, classrooms and restaurants. A 'comfortable' space is often one with many soft surfaces, such as upholstered furniture and carpets, judged as much by acoustics as by vision. Unfortunately, hard surfaces are often cheaper to install and easier to maintain than softer surfaces, so that acoustic problems are more likely to be caused by high reverberation times, that make spaces seem too loud. However, sometimes a highly reverberant space can inspire people to behave respectfully, lowering their voices and walking softly to avoid disturbing the space.[7]

Urban sound

Reverberation times can also be considered for outdoor space. In urban environments, it is common to find spaces that are bounded on each side by the hard, external surfaces of buildings, walls or paving. Even though the 'surface' open to the sky will be absorbent, the reflection from the other surfaces can contain and reverberate sound. Thus courtyards or narrow streets can be lively, or even noisy, as they direct sound into windows of adjacent spaces. A common antidote to the sounds of the street is the acoustic absorption provided by lawns and garden beds, as well as the masking of sound offered by birds or moving water. The quiet of a walled garden provides a dramatic contrast to the noise of an urban environment.

Measuring sound

To appreciate the behaviour of sound in space, it is helpful to understand the units of loudness and their relation to the way sound is experienced. The basic unit of sound, the decibel, is derived from the bel, named in honour of the inventor of the telephone, Alexander Graham Bell. The bel is a logarithmic unit that measures loudness not in a linear scale but as powers of 10, as shown in Table 4.2. An increase of 1 bel indicates that the level has increased by a factor of 10, or an extra zero has been added. Thus a level of 6 bel indicates a sound level of 1×10^6, or 1,000,000 times, the agreed base level. That base level is intended to be the threshold of hearing and is defined at 1×10^{-12} watts per square metre.

Table 4.2 Common sound levels showing comparison of units

	120dB	12 bel	1 watt/m²	1,000,000,000,000 × threshold
	110dB	11 bel	10^{-1} watts/m²	100,000,000,000 × threshold
Factory	100dB	10 bel	10^{-2} watts/m²	10,000,000,000 × threshold
	90dB	9 bel	10^{-3} watts/m²	1,000,000,000 × threshold
Busy street	80dB	8 bel	10^{-4} watts/m²	100,000,000 × threshold
	70dB	7 bel	10^{-5} watts/m²	10,000,000 × threshold
Office space	60dB	6 bel	10^{-6} watts/m²	1,000,000 × threshold
	50dB	5 bel	10^{-7} watts/m²	100,000 × threshold
Living room	40dB	4 bel	10^{-8} watts/m²	10,000 × threshold
	30dB	3 bel	10^{-9} watts/m²	1,000 × threshold
Quiet room	20dB	2 bel	10^{-10} watts/m²	100 × threshold
	10dB	1 bel	10^{-11} watts/m²	10 × threshold
	0dB	0 bel	10^{-12} watts/m²	1 × threshold

With each bel equivalent to an increase by a factor of 10, it is more convenient to use the smaller unit, the decibel (dB), equivalent to one-tenth of a bel. However, the combination of the logarithmic scale and the smaller decibel unit makes this one of the most complicated scales in all of science! Also, since sound occurs at different frequencies, the sound level is usually averaged over the range of sound that can be heard. But since the human ear is more sensitive to mid-range frequencies, around that of normal speech, than it is to sounds of very high or very low frequency, another scale, known as the *'A' weighted scale*, or dB(A), is often used.

Since an increase of 1dB, or one-tenth of a bel, is equivalent to increasing the sound level by $10^{0.1}$, it is necessary to understand what happens to powers of 10 when the exponent is not a whole number. Any integer part, known as the *characteristic*, determines the order of magnitude; that is, whether the number is 10, or 100, or 1000. Any fractional part, known as the *mantissa*, determines where a number lies between these orders of magnitude; that is, whether we are dealing with 200, or 300, or 400, or 786, or any other number, between 100 and 1000. These rarely work out as round numbers on either side of the equation, but fortunately there are three that do: $10^{0.3} = 2$; $10^{0.6} = 4$; and $10^{0.9} = 8$. If we look at some of the values between 20 and 30 dB, for example, they look like this:

30dB =	3.0 bel =	1000 × threshold
29dB =	2.9 bel =	800 × threshold
26dB =	2.6 bel =	400 × threshold
23dB =	2.3 bel =	200 × threshold
20dB =	2.0 bel =	100 × threshold

In other words, while adding 10dB is equivalent to multiplying the sound level by 10: adding 3dB is equivalent to multiplying the sound level by 2. Each additional 3dB will double the sound level again. Thus if the sound level of one person singing were 60dB, two people would have a sound level of 63dB, four people = 66dB, eight people = 69dB, ten people = 70dB and 100 people = 80dB. But by the time large numbers of any particular point source combine together, they cannot act as a single point source because the size of each individual unit means that the sound is spread out across the area of the group.

The key to understanding the decibel scale is to remember that doubling the amount of sound energy will mean an increase in sound level of 3dB; conversely, halving the sound energy will mean a decrease in sound level of 3dB. Similarly, increasing the sound energy by four times will mean an increase of 6dB and reducing the sound energy by one-quarter will mean a decrease of 6dB. As we have seen, sound radiated from a point source will radiate outwards like the surface of a sphere. According to inverse square law, the amount of energy will reduce to one-quarter with every doubling of distance, which results in a reduction of 6dB. In other words, moving away from a sound will only result in a reduction of 6dB every time the distance between source and receiver is doubled.

The difficulty of using a logarithmic scale is further complicated by the complexity of human experience, which depends more on relative than absolute levels. For example, a sound that has twice as much energy (at 3dB higher) is not normally experienced as being twice as loud. When people are asked to identify when a sound has been increased to twice its original level, it usually requires an increase of around 6dB, or four times the energy, for this to occur. One likely explanation is the spatial significance of sound, so that if the distance between source and listener is reduced by half, the sound will seem to the listener to be twice as loud. This may arise from a primitive response mechanism where hearing is used to evaluate the relative threat posed by predators, whose main aim is to reduce the distance between themselves and their victims.

Sound transmission

The best way to stop sound getting from one place to another is by placing a solid wall between them – the heavier the better. The wall needs to be heavy to resist the vibrations of air that cause sound. But the wall also needs to be solid (that is, not have any holes in it) in order to prevent sound getting through the gaps. For a given construction type, it is possible to consult sound transmission tables that describe the reduction in sound level from one side of the wall to the other. A sound transmission index of 30dB, for example, would reduce a sound of 80dB down to 50dB by the time it reached the other side. Even though decibels are logarithmic, the sound level can be subtracted because any wall will reduce the sound by a constant factor, not a linear amount. So a transmission index of 30dB, or 3 bel, means that the sound level will be reduced by a factor of 10^3, or 1000, when compared to the original sound. The reduction or attenuation sound will be different for different frequencies and is usually averaged over a range of 16 values from 125 to 4000 hertz. Lightweight internal walls may achieve around 30–40dB reduction, while masonry walls can achieve reduction of 50dB or more.

Since sound travels in air, any gaps in a wall, or paths around it, may be sufficient to let sound through and thus counteract the effect of the wall. If, for example, there is a hole half as big as the area of the wall, then half of the sound energy will be blocked by the wall, but the rest of the sound energy will pass through the hole. As described above, halving the sound energy will result in a reduction of only 3dB. Thus even if a wall could reduce an 80dB sound to 30dB, the hole will transmit the sound at 77dB. If the hole were only one-quarter the area of the wall, the sound transmitted would still be 74dB; and even if the hole were just one-tenth of the area, the sound transmitted would be 70dB. In other words, the hole will severely compromise the blocking effects of the wall. One of the main strategies for reducing sound transmission between spaces is to plug the holes by sealing air gaps around doors or windows.

Even if there are no holes, sound can travel around obstacles via 'flanking' paths, such as through ceiling spaces or over the top of masonry walls. By locating barriers either close to the source or close to the receiver, the angle of flanking paths can be increased, making it harder for sounds to be transmitted in this way. In internal space, common flanking paths include ceiling cavities and ducts. Like heat, sound can often find weak paths or bridges, passing more easily through studs than through the insulation between. To avoid this, it may be possible to use resilient fixings or to 'decouple' the two sides of the wall from each other. Walls may also have a sympathetic frequency, vibrating more easily at some frequencies than others, so a more effective sound barrier can be made from using elements of different dimensions (for example, stud sizes and spacing or plasterboard thickness) for each side of the wall. Another major source of noise in internal space is from services such as air-conditioners or lift motors, with vibrations transmitted through buildings as structural borne sound. As well as providing sound isolation for plant rooms, it is also necessary to

Table 4.3 Sound transmission loss (dB) for common construction types (after Szokolay)

Construction	Frequency		
	125Hz	500Hz	2000Hz
Plastered double brick wall	41	48	58
Concrete block wall	36	44	55
Concrete slab	35	41	58
Timber floor	18	37	45
Timber floor with insulation	29	39	50
Single-glazed window	17	25	23
Double-glazed window	30	43	47
Plasterboard stud wall	16	38	52
Insulated stud wall	22	46	61
Decoupled stud wall	34	53	57

isolate mechanical equipment using spring fixing. Air-conditioning will also create noise from the movement of air, which is one of the key design parameters for duct sizing, linings and fixings. Since the sound from ducts usually contains an even mix of frequencies, it can often be indistinguishable as background or 'white' noise. This can provide sound masking for spaces such as open-plan offices, which can even be replaced by white noise emitters when natural ventilation is used.

Sound isolation is certainly necessary for theatre and recording spaces for making sound, but may also be useful in areas where privacy is needed, such as hotels, hospitals or dental surgeries. One of the main areas where isolation is needed is multi-residential units or even attached dwellings that share common, or *party*, walls (so called not because of the noise from a party on the other side, but because there are two legal parties involved in ownership of the wall). Party walls often have performance requirements for both sound transmission and fire isolation, neither of which should be readily passed from one dwelling to another. Regulations for sound transmission may also require impact insulation so that footsteps are not conducted to the unit below, preventing the use of timber floors.

Traffic

One of the major forms of sound to be avoided is that caused by motor vehicles along roadways. As we have seen, avoiding sound by doubling the distance between the listener and any point source will only decrease the sound by 6 decibels. With roadways, the combination of multiple, moving point sources effectively makes a linear sound source. From a linear source, sounds propagate in a cylindrical rather than a spherical pattern, where the energy reduces in proportion to the distance. Thus double the distance will mean half the sound energy, or a reduction of only 3 decibels. Since the ability to move away from a roadway is limited by the dimensions of the site, heavy walls, solid construction and intermediate spaces are usually necessary between busy roads and habitable rooms. Creating an acoustic barrier using trees is not viable, since they are permeable to air and will thus transmit sound, although they can have a significant psychological effect. Another option is to reduce the sound emitted by motor vehicles, which has occurred to some extent in recent decades as a result of improvements to combustion engines. An even better prospect is that electric vehicles will become more common, encouraging natural ventilation by reducing the need to seal buildings against the noise and emissions of the street.

chapter five
AIR

I put my air conditioner in backwards. It got cold outside.
The weatherman on TV was confused. 'It was supposed to be hot today.'

Steven Wright

The Earth is wrapped with two fluids: water in the oceans and air in the atmosphere. While our ancestors may have lived underwater, today humans tend to live where land, air and water meet; at the edges of rivers and lakes and oceans, under a layer of air several kilometres thick. The air provides an ideal mix of gases at just the right pressure to sustain human life. Air is a medium from which we draw the oxygen needed for metabolism and into which we expel used oxygen in the form of carbon dioxide. People can breathe most easily within the usual range of atmospheric pressure at or near sea level – around 100 kilopascals. At higher altitudes, pressure decreases, making breathing difficult and possibly resulting in altitude sickness. At lower altitudes – underwater – breathing is possible with scuba tanks, but rapid changes in pressure can result in *barotrauma*, such as the decompression sickness encountered by deep-sea divers who surface too quickly.

Air and buildings

Because of the pressure of the atmosphere, the air we need to breathe usually finds its way into every space of every building, if not through open windows or doors, then through cracks or gaps between the various elements that make up the building envelope. Whether inside or outside of buildings, we are immersed in air. Rarely are buildings so well sealed that they are impermeable to the air around them, with ventilation occurring naturally as air makes its way in and out as a result of movement in the atmosphere. But as with most aspects of the environment, the need to encourage or direct air movement is a qualitative issue: it is important to have the right sort of air in any given space. The issue of air quality arises partly because of its oxygen content, but also because air is a medium that carries many other aspects of the environment, including heat and moisture, smoke and odours, dust and insects, sound and noise. The challenge with any ventilation system is achieving the right balance; opening a window might let fresh air in, but may bring with it dust, noise or other contaminants. Closing a window might keep these out, but result in the build-up of heat or moisture that can make a room feel stuffy and oppressive. Today, the widespread use of 'conditioned' air promises to solve these problems, delivering filtered air in the right amount to the right place at the right temperature. Unfortunately, this ideal is often compromised by air-conditioning systems that are poorly designed or maintained and that minimize access to outside air in order to maximize their own efficiency.

Air and fire

In warm climates, ventilation is an essential part of strategies for cooling by removing heat and allowing evaporation of sweat from the skin. In colder climates, ventilation is necessary to allow fires to burn effectively, removing depleted air and smoke, and enabling the inhabitants to breathe. The need for ventilation originally resulted from the competition for oxygen between the inhabitants and the fire they used to keep warm. Even though the chemistry of combustion was not fully understood until the late 18th century, the ability of fire to *vitiate* or deplete air of its vital forces was readily grasped. Since fires were used for heating, the problem was to limit the amount of warm air lost through the chimney, and thus the amount of colder, outside air that came in to replace it. Many of the developments in fireplace design around the 17th and 18th centuries involved separating the air used by the fire from the warmed air, which could then be circulated throughout a building. Cast-iron stoves slowed the intake of air, allowing it to be heated by the fire and thus improving combustion efficiency.[1] This meant that larger buildings could be heated without having a fireplace in each room, and was particularly useful for buildings such as glasshouses, where fire could damage the plants and dry out the space.

Healthy air

The need for oxygen and the dangers of carbon dioxide were complicated by the emerging science of chemistry and by the still-prevalent *miasma* theory of disease. Prior to the discoveries of Louis Pasteur and his contemporaries in the late 19th century, most disease was thought to be contracted through exposure to foul air, noxious gases or vapours. Windows were often kept closed to prevent exposure to the fetid air that regularly permeated cities before urban sewerage systems were constructed. However, one advantage of the miasmic theory of disease was the development of ventilation systems for hospitals. Designs by architects from the French Academy in the 18th century show patients separated into individual beds, enabling the free flow of air around the body to prevent stagnation and encourage healing.[2] As germ theory became more widely accepted, these principles were applied to other buildings. One of the key aims of Modernist architecture was to design buildings that allowed more light and air into the interior to improve health by removing germs.[3]

Cooling air

Another consequence of experimental science in the 18th century was the greater understanding of the physical processes associated with fluids, and in particular their change of state between liquid and gas. It was found that a reduction of pressure would increase rates of evaporation, which in turn reduced the temperature of the liquid. In the 19th century, these processes were harnessed in a range of mechanical devices that formed the foundation of today's extensive refrigeration and air-conditioning industry.[4] Prior to the development of machines for extracting heat, the main way to provide cooling in amounts suitable for commercial application was by harvesting and transporting ice. Although it had the unfortunate habit of melting in transit, ice was still of great value in manufacturing processes such as brewing lager, or in preserving shipments of perishable goods, especially meat and fish.[5]

Even when mechanical systems were developed, they were often used to create blocks of ice that could then be easily distributed to provide cooling. In buildings, for example, air could be run over ice before being circulated using fans. These systems were typically used in spaces holding large numbers of people, especially theatres and concert halls, where fenestration was limited to avoid disruption. But the build-up of humidity from the ice, and from the people in the space, caused various problems, including a greater risk of condensation. It was not until the work of Willis Carrier in the early 20th century that cooling systems were able to precisely control humidity as well as temperature, and thus avoid the problems associated with water vapour.

Humidity

The relationship between temperature and humidity is a complex one. One of the great benefits of fire in cold climates is that heat will dry out interior space, along with everything in it. As Steven Pyne remarks, 'Although the winter snows drove both humans and free-burning fire indoors, the relentless hearth desiccated the interior landscape like a kiln.'[6] Today, heaters continue to dry out space, while appliances are available for drying hands, hair or clothes using both air and heat. Air is able to absorb moisture in the form of water vapour, and the amount of moisture that air will absorb can be increased by raising its temperature. Unfortunately, the opposite is also true; as air cools, its ability to hold moisture decreases and may even reach a point where vapour in the air returns to liquid and attaches itself to objects. This process, known as *condensation*, is the major problem encountered when cooling air using mechanical means.

The amount of moisture in air is usually described in terms of its *relative humidity*, which is the amount held by the air relative to the total amount it is able to hold at a given temperature, expressed as a percentage. Because the total amount of moisture that the air can hold changes with temperature, raising the temperature of a volume of air will decrease its relative humidity, even though the actual amount of water vapour is unchanged. Similarly, reducing the temperature of air will increase its relative humidity. This measure corresponds with our sensation of humidity, which, like most sensory experience, relates more to relative measures than absolute ones. With low humidity, moisture from the body is easily transferred to the air, and the eyes, mouth and skin may begin to feel dry. With high humidity, the ability to transfer moisture to the air is reduced, limiting heat loss and making the body feel warm and clammy as perspiration builds up on the skin. This is more noticeable at higher temperatures, when the body is attempting to cool itself using evaporation.

Conditioned air

Because cooling air increases its relative humidity, it may be necessary to remove moisture from the air to prevent it from feeling damp. The strategy employed by Willis Carrier in the design of his air-conditioning system was to lower the temperature of air below what was needed until a saturation point was reached that would give the required moisture content, then raise the temperature of the air back up again, thus lowering its relative humidity to a precise and predetermined level. The actual saturation of the air was achieved using sprays, but the overall consequence was generally the removal of moisture at the point of cooling. The process of modifying humidity as well as temperature was particularly useful in industries where accuracy could be affected by expansion and contraction or rates of drying, such as colour printing. Here the added cost of conditioned air was offset by the avoidance of waste and the benefit of reliable production outcomes.[7]

Similar advantages were promoted to owners and tenants of office buildings as air-conditioning began to be introduced in the United States in the late 1920s. Although buildings were fitted with operable windows, ducted air entering from the corridor meant that these no longer needed to be opened; cleaning costs were reduced by keeping out dust, workers were less likely to be distracted by noise from outside, would be more comfortable from the cool air and would have greater control over the air flow. While some air-conditioned buildings were made with sealed external glazing, such as Frank Lloyd Wright's Johnson Wax building in Racine, Wisconsin (1939), it was the Equitable Building in Portland, Oregon (1948), by Pietro Belluschi, that heralded the fully air-conditioned, sealed glass curtain wall office tower.[8] The construction of buildings without operable windows has been commonplace ever since. The relatively low cost of energy and high cost of construction mean that it is usually cheaper and easier to seal external walls and use an air-conditioning system to modify the climate within. However, the increasing concern for sustainable design, through both reduced energy use and improved indoor air quality, is leading to a greater usage of natural ventilation and mixed-mode systems. It is perhaps ironic that the arguments for reducing the reliance on air-conditioning systems, such as improved worker health and productivity, are the same as those made for introducing them half a century ago.

Natural ventilation

The movement of air through buildings is essential in order to provide a regular supply of oxygen to the inhabitants and also to take away the heat, water vapour and carbon dioxide that they produce. Since buildings usually have openings in the form of windows and doors, air will flow through it as a result of the 'natural' movement of air in the atmosphere. When sunlight heats the ground, the ground heats the air and the air expands to create a zone of high pressure. Air will then move from the zone of high pressure to any zones of lower pressure around it, resulting in breeze or wind. When moving air encounters an object, such as a building, the air must move either around it or through it to reach the other side. The physics of air flow is extremely complex, and accurate modelling will often require the use of *computational fluid dynamics* (CFD). However, there are several key factors that determine how much air will move through the building. These are the speed and direction of the wind, the size of the opening through which the air enters the building, the size of the opening through which the air leaves the building and the amount of friction encountered in between resulting from contact with interior surfaces and objects. The quality of air can also be affected by locating windows adjacent to water or green space, which can partially cool and filter the air before it enters a building.

Getting air to flow through buildings is reasonably easy; with openings on either side, and not too much distance between them, the air will move through of its own accord. If central features such as lift cores or corridors block the flow of air, spaces may be left with openings on only one side. Since air cannot easily move in and out of the same opening, ventilation due to natural air movement will only be felt within a few metres of the windows (see Figure 5.1). For spaces with openings on either side, reasonable cross-ventilation will occur for up to about six times the floor-to-ceiling height; about 18 metres for a typical office space (see Figure 5.2). In buildings with larger floorplates, air-conditioning is likely to be required to ensure adequate ventilation.

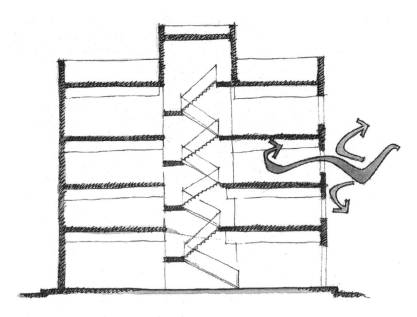

Figure 5.1 Ventilation from one side only

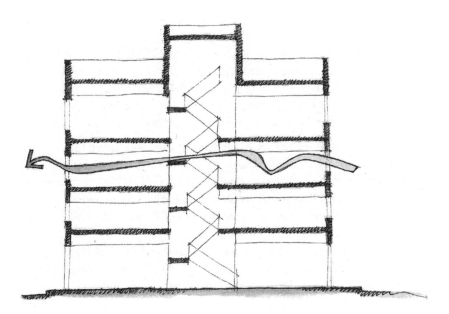

Figure 5.2 Cross-ventilation

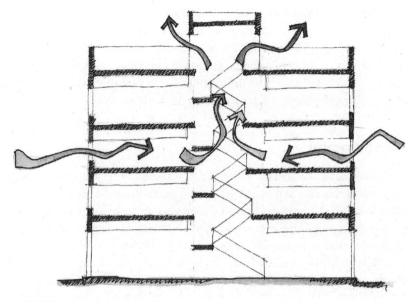

Figure 5.3 Ventilation from stack effect

Rising air

Once inside a building, air tends to absorb heat from people, lights or machines, which causes it to rise. This is known as *buoyancy*, or the stack effect, and can be used to encourage air movement, by allowing air to move upward through atria or other vertical spaces and out through high windows (see Figure 5.3). By using *solar chimneys*, heat from the sun can even be used to create an upward flow of air, which will cause cool air to be drawn in through lower openings. However, using heat from the sun to remove heat through ventilation can be a risky strategy and, if incorrectly designed, the heat gained can be greater than the heat removed, so that occupants, especially those near the solar chimney, can be subjected to an unwanted heat source.

Window design

When dealing with natural ventilation, the question is how much is appropriate: too little, and the air inside can feel stale and oppressive; too much, and the air can move paper or other objects in the space. Unfortunately, this is complicated by the natural fluctuation of air, as it varies in speed and direction, as well as temperature and humidity, every minute of the day and every day of the year. The traditional solution centred around the remarkable invention of the openable window, where glass held in a frame can be pivoted on hinges or moved along tracks to let in air. From a few basic choices comes the rich variety of opening types: hinges on the top, side or bottom; vertical or horizontal sliding; or combinations such as double or bi-fold panels. The type of window will affect how air enters the space, coming from the top, bottom or side of the opening. But it will also affect patterns of vision and physical interaction, determining how people see through a window and whether they are able, for example, to lean their head out of the window to get a better view. Doors provide physical access into space, but windows are the major points of exchange, where air, light and vision penetrate the walls to bring outside and inside together.

Air-conditioning

When the flow of air due to natural ventilation is insufficient, mechanical ventilation can be used to force air into or out of buildings. In domestic environments, fans are often used to avoid condensation by removing air from kitchens, laundries or bathrooms that may contain large amounts of water vapour. Mechanical ventilation is also useful for removing odours from kitchens or bathrooms in larger-scale buildings, or for removing vehicle exhausts from car parks. But the most common strategy is to combine mechanical systems for heating, ventilation and air-conditioning, otherwise known as HVAC. An air-conditioning system usually comprises a 'chiller' unit, designed to remove heat from a fluid (usually water), along with a fan system that forces air over pipes containing the chilled water, then through ducts to the various spaces throughout the building. The cool air then absorbs heat from the various sources around the building, such as people, machines and lighting, as well as the heat entering the building fabric from solar radiation and conduction, before being returned to the chiller unit via return air shafts. The air is then topped up with air from outside to make up for the small amounts lost through cracks or open doors (usually around 10–20 per cent) and then cooled and distributed through the building once again. In other words, fans circulate air in order to remove heat from the building and transfer it to the chiller unit; the chiller unit circulates water to extract heat from the air and transfer it to the outside. In domestic applications, the two components of fan and chiller unit are sometimes separated, in what is known as a *split system*, allowing the noisier chiller unit to be placed away from living areas where the noise will be less disruptive.

With typical air-conditioning systems, the chiller unit used to produce cold water does not actually create *cold*, but simply extracts heat from one object (water or air) and moves it into another (the air outside). But the reason air-conditioning is needed in the first place is that there is no 'cooler' object that can be used to absorb the excess heat. What is remarkable about the chiller unit is that it is able to dispose of heat even when there is no 'cooler' object available – in other words, it works against the thermal gradient by taking heat from one object and moving that heat into another object *at higher temperature*. For this reason, a chiller unit is sometimes known as a *heat pump* because of its ability to push heat 'up' from cold to hot, against the natural tendency or gradient from hot to cold.

It does so through what is known as the *vapour-compression cycle*. This involves the use of a refrigerant, a liquid that will convert to a gas at relatively low temperatures. A common refrigerant, R12, has a boiling point at around 0°C at normal atmospheric pressure, which can be increased to around 40°C at higher pressures (about 1000 kilopascals). By changing the pressure, the refrigerant can be made to boil, and thus absorb heat, at a temperature slightly below our normal comfort level, and then, after being compressed, can be made to condense, and thus release heat, at a temperature slightly above our normal comfort level. The change from liquid to gas or vice versa enables the refrigerant to absorb or release heat without changing temperature, through what is known as the *latent heat of evaporation*.[9]

Thus heat can be extracted from inside air, at a relatively low temperature, and transferred to air outside, even though that air is hotter. Most air-conditioners can also operate in *reverse cycle*, where the refrigerant moves in the opposite direction, extracting heat from the outside air and bringing it inside the building, thus converting the air-conditioner into a heater. Since the heat is being pumped from outside and not created from gas or electricity, this can be reasonably efficient with each unit of energy used to run the pump able to move several units of heat energy. However, since most air-conditioners run on electricity, this does little more than make up for the inefficiencies of combustion at a typical coal-fired power station.

Centralized air-conditioning systems make for remarkably versatile buildings. Combined with artificial lighting, they free space from its reliance upon external walls to provide light, air and thermal comfort for the occupants. In contrast to the narrow floorplates required for daylighting and natural ventilation, air-conditioning allows much larger, 'deep-plan' buildings that can result in greater site coverage and better use of difficult sites. By avoiding the reliance on external walls for climate modification, air-conditioning makes space more flexible, and thus more usable, regardless of the location or climate. Without this versatility, high-density cities, especially in hot climates like Singapore and Dubai, would not be possible.

Equipment and distribution

Space is also needed for 'plant' rooms, which usually take up about 5–10 per cent of the total floor area, so that an average ten-storey building is likely to have one entire floor devoted to mechanical services.[10] The typical location is on the highest floor or roof, which, although valuable space, allows outside air to be drawn in away from vehicle exhausts at street level and also allows cooling towers to operate effectively, dispersing the heat from the building into the surrounding air. The space is often shared with other machinery, especially motor rooms for traction lifts, located at the top of the lift shaft. Hoisting machinery to the top floor can be difficult, but once there, it can be easily maintained, as well as isolated from other levels to prevent noise and vibration from disrupting occupants. With a centralized system, the air intake

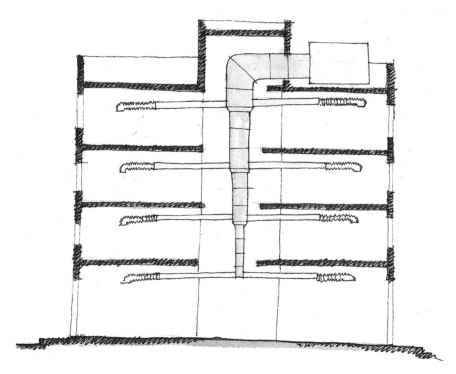

Figure 5.4 Conditioned air distributed through ducts

for an entire building can be isolated at a single location. This avoids the need for operable windows, thus lowering costs for construction and maintenance. It means that the air can be filtered and dried, so that the dust and moisture ends up in the plant room and not scattered throughout the building. It also improves security, preventing people entering through windows, and also preventing objects or people from falling out, inadvertently or otherwise.

From a central location, air is distributed down through vertical shafts, then across through ducts and out through registers or grilles to the spaces throughout the building (see Figure 5.4). Ducts are usually concealed in a ceiling space, hidden behind a grid of suspended ceiling panels that provide a degree of acoustic absorption. Ducts can also be exposed, forming part of an 'industrial' aesthetic that expresses patterns that flow throughout a building. The first major work to do this was the Centre Georges Pompidou in Paris by Renzo Piano and Richard Rogers (1978), in which colour-coded tubes for air, water, electricity and circulation are woven through an open-frame external structure.[11] The design of an air-conditioning system usually involves getting a reasonable distribution of air to each space, taking into account variation in the numbers of people and machinery, as well as external changes such as heat from the sun or shadows cast by adjacent buildings. Since more accurate distribution usually entails greater cost, systems are usually designed to keep occupant 'dissatisfaction' down to a minimum, based upon predicted survey responses. Even the best systems, however, will still return a dissatisfaction rate of around 5 per cent.

Indoor air

In the past half century, air-conditioning has changed from being extremely rare to extremely common. What was once used mainly in theatres and retail outlets to attract customers is now incorporated in nearly every public or civic building, and the number of domestic units is rapidly increasing due to lower costs. One result has been a rise in expected levels of comfort in indoor environments.[12] Another contributing factor is changes to work practices. Many workers are able to tolerate extremes of temperature in their work environment, simply by enduring higher temperatures or using the heat from machinery or physical activity to keep warm. But the increasing number of office workers throughout the 20th century, conforming to strict dress codes and working at fixed locations with hands exposed for typing or other paper-related tasks, has led to a reduced tolerance for temperature variation. One of the key criteria by which leasing agents distinguish quality of office space for establishing rents is the accuracy of the air-conditioning system.

But for all the benefits of air-conditioning systems, there can be problems. Systems that are inadequately maintained or cleaned can perform poorly in terms of filtering the air, and the water removed as condensate at the point of cooling can become contaminated with biological agents, one of the most dangerous being the legionella bacterium, which can prove fatal to building occupants. Other problems arise from the large percentage of air being recirculated, which can be up to 80 per cent in typical systems. The main reason for this is that air is used as a medium for heat, and the cost of heating or cooling outside air is usually much greater than the cost of heating or cooling recirculated air. This works reasonably well when there are no sources of pollutants inside the building, but can be a problem if fittings such as carpet or furniture give off volatile organic compounds (VOCs). Incidences of building-related illness or *sick building syndrome*, where inhabitants suffer a physiological reaction to building contents or systems, are often compounded when the cause of the reaction is kept inside the building by the ventilation system.

However, even when these problems are avoided through better materials specification, it is still possible for inhabitants to feel tired or drowsy when exposed to recirculated air throughout the course of a working day. Also, people will often be more tolerant of variation in temperature resulting from changes in the natural environment than from air-conditioning systems. Studies have found that a broader temperature range is tolerated by those who have a level of control over their thermal environment, such as the ability to adjust air flow rates or to operate a window, than by those who are subjected to a centrally controlled system. Fortunately, employers are realizing that keeping down the costs of the air-conditioning system makes no sense if it makes the people inside less effective at their work. Recent innovations in air-conditioning systems have attempted to address these problems by reducing recirculation rates and increasing user control.

One alternative to traditional ducted air-conditioning is the use of *chilled-beam* cooling. This has been used recently in buildings such as Council House 2 in Melbourne, by Mick Pearce and DesignInc (2006). Chilled beams work like panel heaters in reverse, absorbing radiant heat from people and machines in the space using water circulated to a central cooling system. The panels are usually combined with displacement ventilation, where air enters through inlets at floor level that can be adjusted by occupants, rises naturally as it is heated by bodies and machines in the space, and is vented at ceiling level.[13] The panels, located on the ceiling, also cool some of the warm air that has risen from below, allowing it to drop back down to floor level, where it can absorb heat and then rise again. By using the panels to provide cooling, the air no longer needs to act as a medium for extracting heat, which means that high recirculation rates are no longer necessary and up to 100 per cent outside air can be provided. Also, the occupants experience a combination of radiant and convective heat loss, similar to that experienced in the natural environment.

Another strategy is to use geothermal cooling, drawing air through underground tunnels to cool it before entering a building. In Federation Square, Melbourne, by LAB and Bates Smart (2002), air for the atrium space is drawn in through an underground concrete 'labyrinth' providing passive cooling of up to 12°C in peak summer conditions.[14] Other projects, such as Norman Foster's St Marys Axe tower in London (2004), provide a combination of air-conditioned and naturally ventilated spaces allowing occupants a degree of variety in their thermal environment. Being able to breathe outside air can have a positive psychological effect, as people are able to feel in touch with their environment, and perhaps even to look forward to life beyond work. Recognizing the need for variety in the thermal environment may help to counteract a sense of 'thermal boredom' that can occur when occupants are subjected to the same internal conditions every day of the year.

These recent innovations herald a potential phase change in the built environment. Instead of buildings sealed off from the street to protect from the noise and fumes of motor vehicles, intended more to make the air-conditioning system run efficiently than make the inhabitants work effectively, it may be possible to create more habitable environments inside and outside of buildings and enable them to interact across more permeable building fabric.

chapter six

WATER

I bought some powdered water, but I don't know what to add to it.

Steven Wright

Water and cities

One of the major environmental challenges for the 21st century is to maintain water supplies for increasing populations in the face of declining rainfall levels. This is despite the fact that water networks provide one of the most sustainable forms of consumption in urban environments. Water collected in catchment areas and stored in reservoirs can be distributed to households through a network of pipes using nothing but gravity. Apart from maintaining pipes and protecting catchment areas, little effort is required to ensure an ongoing supply of water – except where annual rainfalls vary from expected levels. Urban water supplies literally 'tap' into the hydrological cycle, the movement of water from the oceans into the atmosphere and back again powered by the sun. Having been lifted into the atmosphere by solar energy, water falls to earth as rain and then makes its way back to the oceans along watercourses such as streams and rivers. Pipes divert the water into buildings, enabling us to control the flow with the simple turn of a tap instead of having to make our way down to a river to obtain water.

Rivers provided the original water supply for many urban settlements, with cities such as London, Paris, Shanghai and New York all founded on navigable waterways that ensured a regular supply of water for drinking as well as for transport. Before urban water supplies were built, access to water meant a visit down to the river, or to the wells where underground water could be brought to the surface. Reticulated water saves the effort of having to carry water to where it is used, enabling cities to grow by allowing everyone to have access to water, not just those who live near natural sources.

But pipes also solve another major problem, which is that people use waterways for disposal as well as supply. Because water *flows*, it carries away much of what is put into it. This is good for the person using it, but not so good for those downstream. As cities grow, the cumulative waste of their populations can transform waterways from clean supply into a polluted drain. The rivers of several large cities have at various times been described as 'open sewers'.

Reticulated water supply prevents the problem of downstream pollution by enclosing water in pipes until it reaches the point of use. In other words, plumbing divides the flow of water from a river into thousands of miniature watercourses that can be brought into homes and reactivated at the turn of a tap. A river can be said to deliver water in *series*, so that one person's waste can become another's supply, while plumbing delivers water in *parallel*, such that each user is guaranteed clean water supply regardless of their location relative to other users. At the point of use, the water moves from a supply network into a disposal network; from water pipes into sewage pipes. From there, the polluted water is brought together and carried to the ocean via treatment plants, or sometimes made to flow into oceans with no treatment at all.

Using water

One of the major tasks of any architect is to anticipate the flow of water in and around buildings. A great deal of effort goes into construction details to prevent the ingress of rain and condensation: flashings, lapped junctions, cavity walls, membranes and seals are all used to keep water out. Then another series of details – fixtures and fittings, wet area linings, water supply and drainage pipes – enable water to be brought in. This allows people to use water in precisely controlled amounts whenever it is needed, without having to carry it, or purify it, or even worry about it at all. In fact, water that is brought into buildings is so easy to use that many people forget what a valuable resource it is, and manage to waste ever-increasing amounts. This is partly because of the myriad uses that emerge when water is readily available, but also because of the tendency to let water *flow* when it is being used rather than simply using the amount that is needed. Advertisements for water conservation might encourage people, for example, to turn off the tap while brushing their teeth. (Meanwhile, ads promote hot-water systems that 'will never run out'.) Although reticulated water supply is sustainable in principle, changing rainfall patterns and increasing populations mean that water supplies are a precious resource. This has prompted governments to undertake advertising campaigns, provide financial incentives and sometimes initiate restrictions to minimize water use.

Incentives for water saving attempt to reverse the trend of increasing water use that has occurred since reticulated supplies were introduced. When French engineers first brought water to the households of Paris, their estimate of daily usage was around 20 litres per person per day.[1] Today, in the United Kingdom, the rate of usage is around

350 litres per person per day. The reason for the difference is that the provision of water to households not only satisfied existing needs, but created new ones. Water is essential for satisfying basic requirements of health and hygiene; cooking and drinking, cleansing the skin and removing bodily wastes. Regular washing is necessary to remove the micro-organisms that cause disease, and its widespread adoption helped to eradicate the epidemics that regularly afflicted European cities up until the 20th century. But beyond the basic need for health, water proved to be wonderfully versatile. It could be used for cleaning clothes and bed linen, for washing interior surfaces, for creating verdant and exotic gardens, as well as for a range of recreational purposes. More water meant cleaner bodies, cleaner clothes and cleaner buildings; social expectations changed and, as a result the amount of housework needed to keep up with those expectations increased.[2] Bathing became a form of relaxation and luxury, allowing the therapeutic functions of water that had previously required a visit to public baths or thermal springs to take place in every home.

Plumbing and hygiene

It is easy to identify great buildings devoted to water, from the Baths of Caracalla in Rome, dating from the 3rd century AD, to the Therme at Vals in Switzerland by Peter Zumthor (1996). But a more significant impact on architecture can be seen from the introduction of water into domestic space. The early years of architecture's Modern movement coincided with the discoveries of scientists such as Louis Pasteur and Robert Koch, who during the 1890s, identified germs as the source of disease. Many architects joined the campaign for urban reform, seeing the new practices of medicine and hygiene as being sympathetic with the ideals of Modernism.[3] Many of the early experiments in functional architecture occurred with hospital design, such as the Sanatorium at Paimio in Finland by Alvar Aalto (1932). Adolf Loos wrote an article about plumbing, lamenting the poor bathing practices of his fellow Europeans and equating bathing with culture.[4] The preference for 'clean' lines, white walls and transparent surfaces of glass, fundamental to Modernist architecture, were heavily influenced by the aesthetics of the hospital.[5]

Le Corbusier was a particularly ardent promoter of hygiene. When he wrote that 'a house is a machine for living in', he coined one of the most recognizable metaphors of the Modern movement. But the functions he had in mind for the machine house are not entirely technological. He wrote: 'A house is a machine for living in. Baths, sun, hot-water, cold-water, warmth at will, conservation of food, hygiene, beauty in the sense of good proportion.'[6] His manual of the dwelling in *Towards a New Architecture* reads like the advice of a sanitary reformer:

> Demand a bathroom looking south, one of the largest rooms in the house or flat, the old
> drawing-room for instance. One wall to be entirely glazed, opening if possible on to a
> balcony for sun baths; the most up-to-date fittings with a shower-bath and gymnastic

appliances ... Never undress in your bedroom. It is not a clean thing to do and makes the room horribly untidy ... Demand bare walls in your bedroom, your living room and your dining-room ... Teach your children that a house is only habitable when it is full of light and air, and when the floors and walls are clear. To keep your floors in order, eliminate heavy furniture and thick carpets.[7]

Along with changes to the external appearance (and internal finishes) of buildings, plumbing changed the very kind of buildings that were able to be built. Prior to the introduction of flush toilets, most houses required a connection to the ground that enabled a pit toilet to be built, usually in a rear yard. The urban pattern of terraced housing with a rear yard and access lane was designed to enable a 'night cart' to remove the waste and sell it as fertilizer. Even the residents of walk-up tenement blocks tended to share a toilet dug in common space at the rear. Plumbing made this type of toilet obsolete, transforming rear yards from a place of waste to a place of recreation. But it also meant that dwellings no longer needed to be built at or near ground level. The residential towers proposed by Le Corbusier for his Radiant City, like all high-rise buildings, were only possible because of plumbing. Elevators and steel-frame construction made towers possible, but plumbing made them *habitable*.

Plumbing and privacy

The disconnection from ground also means that spaces for hygiene can be wholly contained within a dwelling. The relation between water and nudity means that spaces for bathing are usually the most private part of any dwelling. Many fixtures were originally installed in storage spaces; hence the term *water closet*. When located along an external wall, bathing spaces usually have small, high windows, often with opaque glass, to prevent the occupant from being seen. But mechanical ventilation and plumbing also allow bathrooms to be built without windows, enabling them to be located centrally in deep floorplates, thus freeing the external skin for less private tasks. A fully glazed facade, such as that used for the Farnsworth House by Mies van der Rohe in 1950, is only possible when ablution spaces are wholly internalized. However, even with the bathroom spaces shielded from view, the level of privacy enjoyed by Mrs Farnsworth was less than ideal.[8]

Bathing rituals

Today, the bathroom is the site of a complex array of strategies for cleaning and grooming, including washing of hands, face, teeth, hair and body; the application of lotions, make-up and perfumes; shaving, brushing, waxing, cutting or treating hair and nails and, possibly, storage and dispensing of medicines. Warm water can also be used to relax or revitalize the body, or simply for sensory pleasure. These acts are made possible by the fixtures and fittings of the bathroom, such as bath, shower, basin and, possibly, bidet.[9] Each provides a flow of water into which the body can be placed, in whole or in part, for the purposes of cleaning, and is also designed to collect and remove the water after it is used. Practices of domestic ablution reveal little of the former, sacred status of water and its use in religious ceremonies for cleansing the body in order to achieve purity of the soul.[10] The modern bathroom might be seen as having secularized the rituals of cleansing. For many people, a daily bath or shower is an informal ritual, necessary not so much for removing dirt, but to signify preparation for, or completion of, the working day.

When workers are employed to undertake physical labour, washing may involve removing dirt encountered at work in order to prepare for social activities with family or friends in the evening. But for office workers, the major source of 'dirt' is the body itself, whose excretions and odours must be removed in order to be socially acceptable in what is typically a high-density, air-conditioned environment. Thanks to the protection of clothing and the relative cleanliness of most urban space, there is very little dirt that manages to accumulate on the skin. Instead, the ritual of bathing is used for waking up or for winding down, or as an essential preparation for social interaction.

Because of its role in maintaining both the life and lifestyle of urban populations, it is easy to consider water, and the way it is used, as 'natural' phenomena. But as Jean-Pierre Goubert has suggested, the construction of reticulated supply transformed water from a gift of nature into an industrial product.[11] It is possible, for example, to use water from a tap – collected, treated, distributed and heated – at the same time that its more 'natural' counterpart falls on the roof above your head and is diverted to stormwater drains below. The water from a tap is qualitatively different from other forms, since, for the most part, it can be relied upon to be safe. Combined with the practice of locating pipes underground to protect them and ensure gravity feed, this tends to conceal the origins of water and thus conceal its importance to urban life. Water is a largely invisible product without which cities as we know them would not be possible. Its daily use by the inhabitants of a city, for cleaning bodies, clothes and building surfaces, occurs only because of acquired and transmitted practices. One of the first tasks of any parent is to train children in basic hygiene; to use toilets, to wash hands and to shower or bathe on a regular basis. When water supply was first introduced, this sort of frequent use was not automatically adopted, but needed to be promoted and disseminated by sanitary reformers and hygiene movements.[12] What began as a means of promoting health and preventing disease has transformed into an essential aspect of urban life.

Public convenience

Water is often celebrated in public spaces, with fountains or water features, as well as parks and gardens with non-native species, demonstrating a ready availability of water for both residents and visitors. Fountains act to cool the air in public spaces and were once thought to purify the air and help prevent disease.[13] Originally connected to wells or standpipes, fountains have been rendered obsolete as a source of water for urban populations, although drinking fountains are often provided in major public spaces. Access to water may also be provided in the form of public toilets, although this may be motivated less by hospitality than by the need to discourage men from urinating in alleyways, doorways or city streets. Unfortunately, public toilets may be more of a risk than a convenience, instead providing a venue for visitors to take part in various illicit activities.

Water supply

Utilities such as water supply and electricity are the most convenient of all consumer goods. With supply pipes brought directly to the point of consumption, there is no effort involved in collecting the goods that they provide and no need to plan for patterns of use. Instead, the turn of a tap or flick of a switch enables them to be used whenever they are needed, in whatever amount. Payment is made retrospectively, with a meter, usually installed by the supply authority, used to record how much has been used. Utilities can be used in this way because of the fixed location of the consumer. Having invested in infrastructure of pipes and meters, suppliers can then benefit from ongoing usage with little risk of theft or lack of payment. Most of the risk is borne by the consumer, who is susceptible to problems such as interruptions to supply or excessive cost increases. Equitable and affordable access to reliable supply was initially guaranteed by government ownership of utilities. However, in recent years, there has been a trend towards privatization, with treatment regulated by governments through codes and standards.

Treatment ensures that water is *potable*, or suitable for human consumption, by removal of pathogens, as well as providing a clear and colourless appearance with an absence of odour. Supply is also regulated to ensure reliability, adequate flow rate and pressure. This usually involves collection of water in protected catchment areas, storage in dams or reservoirs, and filtration and treatment before distribution through underground pipes. Treatment may include removal of mineral content (especially calcium carbonate) that makes water feel *hard* and difficult to form lather with soap products. It may also include the addition of fluoride to boost natural levels, which helps to prevent tooth decay. Water pressure is ensured by pumping or gravity, with supply pressure usually the equivalent of about 10–50 metres of *head*, or the pressure that you would get from a water tank that high in metres above the ground at the point of use. This means that a good flow rate can be achieved using only small-diameter pipes, with domestic supply pipes usually 12–25 millimetres in diameter. The pressure means that there is no need to pump the water into the tap, no waiting while the pipe fills up and no need to flush out the pipes to get rid of stale or contaminated water. A tap can simply be turned on to release water that is clean, clear and ready to use.

With water available in pressurized, small-diameter pipes made of copper or hard plastics, there is virtually no constraint over where a tap can be located in any building. The pipes can be easily concealed within the structural thickness of walls, floor or roof space, and can be run in any direction without affecting supply pressure. Only with high-rise buildings, in which the head of water available is insufficient to reach the upper floors, is it necessary to pump water to a supply tank at the top of the building. The architect's task is simply to identify locations where water will need to be used and to specify suitable fixtures and fittings. These may include a kitchen sink, large enough for hands, dishes and preparing food; in the bathroom, a basin designed for cleaning hands, face and teeth; a shower or bath large enough to wash the whole body in either vertical or horizontal position and a toilet for depositing and removing bodily waste; in the laundry, a tub big enough for clothes or sheets, as well as taps linked directly to a washing machine; in the garden, taps for watering plants and possibly direct connections to swimming pools or sprinkler systems. Wet areas must also be lined with waterproof materials to prevent splashes and condensation from seeping into walls and floors where it can cause structural damage. Other considerations include the design of storage and bench space, selection of tiles and partitions, mirrors and lighting, heaters and exhaust fans, racks, rails and hooks for towels, toilet roll holders, soap trays or dispensers, hand dryers, paper towels and waste bins.

Waste disposal

Each of the fittings described above is designed not only to accommodate the objects or part of the body being washed, but to collect the water and channel it into a drainage system. Like water supply, sanitary disposal is heavily regulated, intended to protect health by preventing contamination or infection. Disposal systems are largely invisible, hiding the waste in underground pipes, masking the odour that arises from it and, where possible, preventing it from reappearing. Like water supply, waste disposal is gravity-fed, except that since the flow of the water has been interrupted, it must be started again on its downward path. This means that disposal pipes must be located below the point of use, generally built into the floor, and must be laid at an angle to ensure that waste water travels down and away from the building. Drainage pipes must also be larger than inlet pipes to allow for the addition of waste matter. A kitchen or bathroom fitting may drain into a pipe with a diameter of only 45–65 millimetres, but a toilet is typically fitted with a 100 millimetre diameter pipe (see Figure 6.1). Small pipes must feed into larger pipes, so that they increase in diameter away from the drain. Space is required in footings or underground trenches to accommodate these pipes, and in multi-storey buildings, drainage must be brought down to ground level in a vertical shaft, or *stack* (see Figure 6.2).

Figure 6.1 Typical drainage layout for a bathroom

Where water enters the drain, outlet pipes must be open to accept the waste poured into them, but also closed to prevent odours from entering. This is usually achieved using an *S trap*, which allows a section of water to remain in the pipe to create a seal. When drainage systems were first built, vent pipes were attached to allow the escape of sewage gases, which were thought to be a source of disease. While odours are no longer considered dangerous, vents are still used to balance air pressure in the drain, preventing water in the trap from being blown out. This is usually done using a *head vent*, installed at the head or highest part of the drainage run. The vent is a small-diameter pipe connected to the sewerage line at one end, with the other end left open, lifted above the roof-line to avoid odours entering the house and capped with mesh to prevent the entry of insects or other animals.

The design of drainage networks must also take into account the possibility of blockage. With outlet pipes sealed only by a layer of water, it is possible for blockages to redirect waste back into the building. To prevent this, a *flood gully*, or large opening, can be installed outside the building at the lowest part of the drainage run – below the level of all other traps. If a blockage occurs, it is helpful to be able either to dislodge it or to identify precisely where it is to enable that part of the pipe to be dug up and replaced. This is facilitated by the use of inspection openings, where a section of pipe rises to

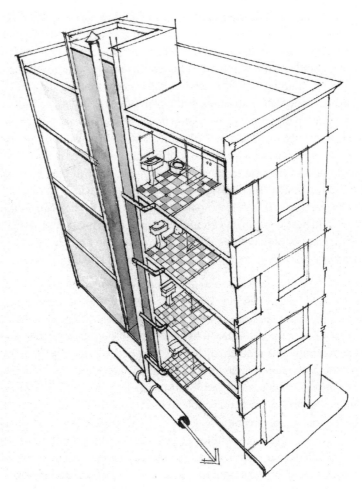

Figure 6.2 Wet areas linked vertically using a stack

the surface, with a removable cap enabling a corkscrew cable to be threaded into the drain. Inspection openings are usually installed where toilets join the main drainage run, and at changes of direction, as this is where blockages are most likely to occur. They are also installed at 30-metre intervals along a straight run, to avoid pipes being too long for the cable to reach through.

Hot-water systems

Since water is brought into buildings through underground pipes, it enters at the same temperature as the ground, which is generally below air temperature and thus cool or cold to the touch. For some uses, especially cooking, water can be heated to boiling temperature using a kettle or a pot placed on a stove top. For many other uses, especially bathing and washing clothes, it is helpful to have a regular supply of 'hot' water, below boiling temperature to avoid scalding, but above cold water temperature to improve comfort and to clean more effectively. This can be done by heating water on a stove as and when it is needed, although this requires a little effort and patience. Instead, most buildings are fitted with a hot-water system to allow heated water to be available, along with cold water, at the turn of a tap. Water can be preheated and kept in a large tank ready for use, known as a *storage system*. Alternatively, water can be heated only when it is needed, which is done using an *instantaneous system*. Both types of system suffer some loss of heat. In a storage tank, the water is able to absorb most of the heat emitted by the element, but will then gradually dissipate heat to its surrounds. This is known as *stand-by loss*, which can be minimized by insulating the tank. With an instantaneous system, water is heated as it flows through the pipes, and heat is lost in and around the pipes, ensuring that the water reaches the right temperature. Both types of system are regulated to provide water at 50°C – only slightly above body temperature – in order to prevent scalding.

Deciding which system will lose the least energy depends on how much, and how often, hot water is used. Storage systems are ideal for high-volume, regular use, such as for a family using the same amount every day. When the entire tank is used, it may take several hours for hot water to become available again, although *quick-recovery systems* overcome this problem by heating part of the water at a time. Instantaneous systems are more suited to small-scale or intermittent use. The choice of fuel type depends upon available supply, with most systems powered by gas or electricity. In environmental terms, gas is preferable to electricity from coal-fired power stations, which produce more greenhouse gases for the same amount of energy. Storage systems have the advantage of linking to cheaper *off-peak* (night tariff) electricity or to roof-mounted solar collectors. Although solar panels usually involve greater up-front costs, they reduce ongoing fuel costs and cause no greenhouse gas emissions. In many places, the increased costs can be offset by government rebates intended to encourage environmental benefit.

The size and location of a hot-water system depends on whether a storage or instantaneous system is used, and also on the fuel type. The original practice of placing hot-water tanks in a roof space was done to allow gravity feed to the taps below. Since tanks eventually corrode, this often meant houses were flooded whenever a system wore out. Today, systems are typically mains pressure, which allows them to be located in more accessible spaces. Instantaneous systems consist of a small heating unit, less than half a metre in each dimension and usually wall-mounted. The size of a storage system depends on the tank, which can be up to several hundred litres, usually made 'upright' so that they can be located in a storage cupboard. Each system can be located indoors or out, although gas systems of either type need to be vented to the outside. When solar panels are used, they can be linked to a floor-mounted tank or to a roof-mounted (*close-coupled*) tank connected to the panels. An extra consideration for solar panels is that they need to face the part of the sky from which there is likely to be the most sun. This usually means facing the panel towards the position of the sun at noon solar time during either equinox; in other words, the middle point of the solar chart for that particular latitude (see Chapter 1).

One further consideration for locating a hot-water system is to minimize the waste of water resulting from what is known as *cold run*. Whenever hot water is required, the water in the length of pipe from the hot-water service to the point of use must be 'run off' until the hot water reaches the tap. This not only wastes water, but also energy, as the heated water remaining in the pipe is then left to cool. This can be minimized by grouping wet areas together and locating the hot-water service centrally between points of use (or slightly closer to those used more frequently). It can also be prevented, but only with added infrastructure, with pumps or ballast tanks now available to reuse the water from the cold run.

Remote systems

Where no mains systems are available for water supply or waste disposal, it may be necessary to use remote systems, especially rainwater and septic tanks, for on-site collection and disposal.

On-site water collection

Rainwater tanks can be used to store water collected from roofs, with sizes depending on roof catchment area and local annual rainfall, as well as on usage requirements. Due to the unpredictable nature of rain, tanks must be sized to allow for dry periods, since a tank large enough for a week's supply will be insufficient if no rain falls for a month. Tank water also needs to be protected from contamination. Dirt can accumulate in gutters between rainfall periods, which can be prevented from entering the tank using *first-flush diverters*. Tank water should also be protected from insects, since the still water provides ideal breeding conditions for mosquito larvae. Tank sizing, along with dimensions of gutters and downpipes, can usually be determined using manufacturers' data sheets, and roof pitches designed to direct water to the tanks. In urban locations, rainwater can be used to reduce reliance on mains water, used, for example, for flushing toilets or watering gardens.

On-site waste disposal

Septic systems combine a storage tank for waste water, allowing the settlement and eventual removal of solids, with seepage trenches to disperse the water into nearby soil. A domestic system typically comprises an underground concrete tank of several cubic metres, with a seepage field of around 200 square metres. If this much land is not available, or if the soil has insufficient absorption, or if it is too close to waterways, an *aerobic waste treatment system* (AWTS) may be required. An AWTS connects the separation tank to a secondary tank, where air is pumped through the waste water to encourage aerobic decomposition. The waste water is then used to water plants via surface sprinklers.

Another option for on-site waste disposal is the use of *compost toilets*. Instead of flushing away bodily wastes with water, compost toilets store the waste in a large chamber where air is circulated with fans to encourage breakdown by aerobic bacteria. Once it is broken down, the waste can be removed for use as fertilizer. The main advantage of this type of toilet is that it makes water from other outlets immediately usable as *grey water*, since it no longer contains human waste. Although the odour produced by aerobic breakdown is far less offensive than the anaerobic breakdown typical of water-based systems, that odour is more evident to the user without the masking effect of water.

On-site collection and treatment can contribute to sustainability by reducing the demand for mains water supply and waste disposal, but require localized construction and maintenance of systems as well as sufficient space to allow them to operate effectively. Networks of water supply and waste disposal separate the land needed for collection and treatment from site of use, avoiding problems such as odours and making regulation easier. But in doing so, they serve to conceal the environmental impacts of water use. In contrast, localized systems provide a more visible demonstration of the cycles of water upon which we depend for health and hygiene, and thus have the potential to encourage better practices of water use.

Water saving

In recent years, population growth and changes in annual rainfall have highlighted the limited capacity of water supply around the world. Many water-saving features, such as dual-flush toilets, are widely available, while governments have also introduced restrictions to limit water use and minimize wastage. Hosepipe bans, for example, slow down the rate of water usage on gardens by forcing people to carry water from taps to where it is needed. Another change that is likely to take place involves reuse and recycling of water, at both domestic and urban scales. Currently, all the water supplied to households is made suitable for human consumption, even though only a small amount is actually used in this way. One option is to provide a second pipe to all households, allowing one for potable and one for non-potable uses. This may be a good option for new settlements, but the cost of retrofitting in existing situations is prohibitive. Another option is to allow water to be used for more than one application depending on the level of purity required. Recent designs for a toilet cistern, for example, collects water from the handbasin to use for the next flush. Another version is the use of grey-water systems, allowing water from baths, showers and basins to be used to water gardens instead of being directed into the sewage system. Although such systems require more complex infrastructure, they are likely to become more prevalent as costs increase and existing water supplies are made to serve the needs of increasing numbers of people. Another source of water supply is desalination, converting sea water to drinkable water by removing the salt content. This removes the dependence on rainfall, making water supply more reliable, although at a cost of increased energy use. To counter this, many of the plants currently being developed will be powered by renewable energy sources such as wind turbines.

chapter seven

FIRE

I used to work in a fire hydrant factory. You couldn't park anywhere near the place.

Steven Wright

Fire is essential for buildings, for creating materials such as concrete, glass and steel, for construction processes such as welding and brazing, and for creating electricity to power air-conditioners, artificial lights and other appliances. Harnessing fire in this way means that it is no longer visible and is instead hidden away in distant power stations or locked up in objects in the form of embodied energy. Before it could be converted to electricity, fire was used extensively in urban environments, providing heat, light and power in almost every house and factory. When it was more prevalent, it also managed to escape control more often, leading to the great fires that ravaged cities throughout the world, from Rome in AD64, to London in 1666, to Chicago in 1871. These fires, though tragic, did have some benefit, often prompting urban regeneration and helping to build the careers of aspiring architects. The fire of London, for example, allowed Christopher Wren to build nearly 50 new churches, including St Paul's Cathedral; in the years following the great fire of Chicago, the building boom allowed architects such as Louis Sullivan and Daniel Burnham to build their careers designing simplified high-rise buildings in what came to be known as the Chicago School style. Lessons from these fires led to the various regulations still in place today, aimed at preventing fires and limiting the damage they cause.

The danger of fire has even been used by Hollywood moviemakers to scare audiences. In Irwin Allen's 1974 film *The Towering Inferno*, guests at the opening-night party to celebrate the completion of a new 135-storey building in San Francisco find themselves trapped by fire, which started as a result of poor-quality electrical wiring. As the fire breaks out, the building's architect, played by Paul Newman, discovers the wiring problem and must then join the battle to save lives and extinguish the fire. In one notable scene, Newman requests a list of tenants in the new building, explaining that he needs to know what they were doing in the building to judge the effect of the flames; a ping-pong ball manufacturer, he wisely claims, would lead to cyanide gas if engulfed by fire. Just what a ping-pong ball factory would be doing in a skyscraper is not explained, but the scene establishes the architect's intimate knowledge of fire and its consequences for the building he has designed.[1] *The Towering Inferno* was one of a series of 'disaster' films made in the 1970s, such as *Earthquake* and *The Poseidon Adventure*. The disaster film emerges from the genre of the horror film, where the danger of monsters such as vampires or werewolves is replaced by the threat of natural disaster. This time, the danger is caused by fire, which has the capacity to turn monstrous when it escapes control. But the danger of fire was only part of the story: the film was also a warning about the hubris of building skyscrapers, a modern-day version of the Tower of Babel.

The behaviour of fire

Fire is a process of combustion, oxidizing raw materials and creating heat, smoke and gases, each of which can be dangerous for humans. The heat from fire, readily reaching several hundred degrees, is far beyond what the human body can tolerate, with firefighters needing special protective equipment for temperatures above about 50°C. That heat can result in horrific and extremely painful burns, or quickly lead to death. By consuming oxygen and replacing it with carbon dioxide, or poisonous gases carbon monoxide or hydrogen cyanide, fires can suffocate or poison inhabitants long before they come into contact with the heat or flames. As a result of incomplete combustion, fires will also produce smoke, which can limit visibility and make it difficult to breathe, preventing occupants from finding escape routes and hindering firefighters from conducting search and rescue, and extinguishing fires. Significant danger also occurs from the threat of explosion or the rapid spread of fire. Even when highly volatile materials are not present, a fire within a building can slowly heat surfaces such as walls and ceilings until they reach combustion temperature, or *flashpoint*, when they suddenly ignite causing the room to fill with flames. This rapid spread of fire throughout a space can be extremely dangerous, especially as it can mark the point where the building itself is on fire and not just the objects within.

While fires can destroy buildings, it is the threat they pose to human life that is of greatest concern. When fire does escape control, it becomes an unwanted presence and steps must be taken to evacuate the people or extinguish the fire, or both. Unfortunately, fires tend to move through buildings along the circulation paths designed for people: up staircases, along corridors, in and around rooms. Fires thrive in the protection provided by buildings and readily consume the dry and often combustible materials they contain. In doing so, fires negate the protection that buildings provide for people, transforming a building from a place of refuge to a place of danger. When a fire moves up and through a building, the people inside must move in the opposite direction, down and out, away from the danger presented by the fire.

When invaded by fire, a building must suspend the functions needed for everyday use and instead contain and limit the spread of fire and enable people to evacuate as quickly as possible. The spread of fire is usually so rapid that this must happen instantaneously; the provisions for containment, extinction and evacuation must be already in place if and when a fire does occur. Of course, the ideal fire prevention system is one that never needs to be used. One of the reasons that fires are so rare is precisely because of the success of fire-prevention strategies and regulations that do exist. Today, building fires are an extremely rare event, given the number of buildings that exist throughout the world. The only unfortunate consequence is that many people are unfamiliar with the experience of fire, so that when tragedies do occur, it is often because building owners or operators fail to meet fire-safety regulations or because people fail to follow basic safety procedures.

The behaviour of people

Because of the threat posed by fire to human life, it is generally hoped that people will avoid or prevent fire from occurring, will behave reasonably in its presence and will evacuate any building in which it occurs in an orderly manner. In most cases, people do not experience a fire directly, but are alerted to its presence via an alarm or other warning device. Fire alarms can be met with a bored indifference, their sound taken to mean malfunction or the inconvenience of a drill rather than the actual presence of a fire. Appointed wardens are usually required to ensure that people take alarms seriously, whether for the purposes of a drill or in the case of an actual event. Even when a fire occurs, people will often try to collect valued possessions and, in some cases, have been known to run back into buildings to collect them. But the main danger occurs from failing to obey fire regulations. The most common problem is entertainment venues such as nightclubs or discos held in spaces not designed for large numbers of people, or where means of egress are not suitably maintained. Many tragedies have occurred because fire exits have been blocked to prevent people from entering through side doors, limiting the options for escape when fire does break out.[2]

One tragic example is the fire at the Triangle Shirt factory in New York in 1911. The factory occupied the top three floors of the ten-storey Asch building and was staffed mainly by European immigrant women making garments. When a fire broke out among the textiles on the eighth floor, people on that floor managed to escape and a warning was sent to managers on the top floor. However, workers on the ninth floor received no warning and the fire was well developed before they became aware of it. By that time, the main stair was filled with smoke and flames, and the other door, blocked to prevent workers from taking breaks or stealing fabric, could not be used. The fire caused the elevator to stop working and the single external fire escape collapsed under the weight of those trying to escape. Workers waited to be rescued from upper windows, only to find that ladders used to fight the fire only reached the seventh floor. Although the fire only lasted for about 30 minutes, the final death toll was 146 out of only 500 employees. The Triangle fire was the catalyst for extensive changes to both fire laws and labour regulations throughout the United States.[3]

Detection

The processes of evacuating people and minimizing damage need to start as soon as a fire is discovered. While it is hoped that fire will be seen by people within a building, it is also necessary to install detection systems in case a fire occurs when people are asleep or where there are no people present. Detectors usually respond to fire-related phenomena, such as smoke or excessive heat, since the actual detection of flames is only likely to occur once the fire is well advanced. Smoke detectors, for example, are now required in all commercial and residential buildings.

Communication

Once a fire has been identified, the next task is to communicate this to people in the building so that evacuation can take place, and possibly to firefighting authorities so that they can attend the fire. An alarm system can alert people within a building, and telephone lines can be used to notify fire authorities; in larger buildings, a fire indicator panel must be connected to alarms throughout the building to enable firefighters to identify the exact location of the fire.

Evacuation

Once people are alerted to the presence of fire, evacuation should then be possible via escape routes either to a point of refuge inside the building or to an assembly area outside the building. To assist navigation in event of a fire, escape routes are often indicated on reproductions of building plans attached to a wall. Since these are usually printed according to the architectural convention of north facing up the page regardless of which direction the viewer may be facing, they may require a degree of interpretation. For those who cannot read plans, more direct indications of escape routes must be provided in the form of exit signs that light up using either mains or battery power to be visible even when smoke is present.

Exit routes from multi-storey buildings are typically via fire-isolated stairwells, which must lead directly to the building exterior. The dimensions of fire stairs are sufficient to allow traffic in both directions – people moving down and out and, if necessary, firefighters moving up and in. For people who lack mobility, the stairs may act as a point of refuge to wait for assistance from firefighters, or fire-isolated rooms may be provided adjacent to stairwells for the same purpose. The number of fire-isolated stairwells in any building is determined by 'distance of travel rules', which specify the maximum distance of travel from any point in a building to a single stair or to a point of choice between two stairs. The point-of-choice rule anticipates that, even if fire is blocking the path of escape in one direction, escape will still be possible in the other direction. Distances are specified in building codes, and depend on the *class* or type of building, whether commercial, residential, and so on. Since the distance of travel can be in any direction, fire stairs are often included as part of a central core containing lifts and wet areas, which must then be connected to the exterior at ground level.

Isolation

Delaying the spread of fire is necessary to give people sufficient time to escape, which is also facilitated by preventing building collapse. Fire-resistant materials such as concrete and brick are commonly used, but steel structures, which can rapidly deform when subjected to excessive heat, must usually be coated in concrete, plasterboard or fire-resistant coatings. On external facades, fire-resistant spandrel panels or projecting flame barriers are used to prevent transfer between floors or to adjacent structures. In buildings of larger floor area, lateral spread may be prevented by compartmentation, dividing the building into smaller zones separated by self-closing, fire-resistant doors. Vertical spread is usually avoided by limiting vertical connections such as open staircases to only two or three floors, as well as isolating and fitting self-closing fire doors to any vertical shafts.

For many years, fire regulations have encouraged the typical central-core, repetitive floorplate, glass-clad tower that dominates most cities. In recent years, a shift to more 'open' style work spaces has required innovative solutions that still conform to fire regulations. In the Campus MLC refurbishment in Sydney by Bligh Voller Nield, completed in 2000, two sets of staircases were cut into the floorplates at each level. Although the appearance of vertical connection is maintained, fire isolation has been achieved by enclosing every second stair in fire-resistant glass.

Extinction

The final stage of dealing with fire is extinction, or attempting to put the fire out. The method of extinction depends on the type of fire, whether fuelled by fibrous material, flammable liquids, flammable gases, combustible metals, electrical hazards or cooking oils. Methods available include hand-held extinguishers, manual hydrants and sprinklers, as well as provision for specialist firefighting techniques. Hand-held extinguishers are available in various types, including water, foam, carbon dioxide, dry chemical and wet chemical, distinguished by different colours and labels. These generally contain about 9 litres of water, powder, compressed gas or other liquids, depending on nearby fire hazards. Since hand-held extinguishers have a discharge time of only around one minute, they are usually intended to assist evacuation or to extinguish fires at early stages. For more developed fires, a built-in fire hydrant/fire hose reel (FH/FHR) can be used, installed near fire escape stairs to assist clearing a path of escape. These usually consist of a small-diameter hose about 18 metres long that can be operated by building users. Larger diameter hoses are generally brought by firefighters when needed, and attached to hydrant risers located within fire stairs.

Sprinkler systems

Perhaps the most effective form of firefighting equipment is provided by automatic sprinkler systems. Sprinkler heads located at regular intervals, usually every 12 square metres of floor area, mean that a source of water is instantly available at every point within a building. Sprinkler heads consist of a seal that is broken by heat above about 70°C, discharging a stream of water in a radial pattern on the floor below. The slight drop in pressure caused by a sprinkler head discharging will also act as a detection device when connected to other fire systems. Where water presents a hazard to building contents, such as with electrical installations or valuable items in libraries or art galleries, sprinkler systems can be installed using foam, dry chemical or carbon dioxide to minimize damage. Sprinkler systems are extremely effective, with many fires extinguished by water from only a few sprinkler heads before they are able to develop. Although expensive, sprinklers are essential for many building types; and are often mandatory for buildings more than a few storeys in height, where fires cannot be easily reached using ladders and other firefighting equipment. Building codes contain detailed requirements for the installation and maintenance of sprinkler systems, along with necessary provisions for detection, communication, evacuation, isolation and extinction.

chapter eight

ECOLOGICAL DESIGN

A friend of mine once sent me a postcard with a picture of the entire planet Earth taken from space. On the back it said, 'Wish you were here.'

Steven Wright

The flow of energy, air and water play an essential role in architecture, bringing buildings to life by making them warm or cool, bright or dark, loud or quiet. By considering the patterns of flow of each of these elements, architects can transform buildings from inanimate objects into habitable spaces, responding to the needs of the people within, and allowing them to connect with the environment outside. In recent years, a growing environmental movement has inspired renewed concern for energy, air and water, and how they can be used more efficiently and effectively to improve the environmental performance of buildings. But these elements have always played a significant role in architecture. They affect function by providing a suitable internal environment for the various activities that occur within buildings and they affect aesthetics by determining the way buildings interact with human sensory experience.

The environmental, functional and aesthetic aspects of architecture are always interconnected, affecting elemental decisions of form and fabric that determine the level of interaction between internal spaces and the external environment. By relating the needs of the occupants within to the surrounding conditions of climate and site, buildings can be designed to filter the external environment, capturing available sunshine or cooling breezes, or protecting from extremes of heat and cold, rain and wind. For small-scale buildings in suburban or remote locations, these issues are easily resolved, with doors and windows bringing daylight and views to each room, and able to be open or closed depending on the weather outside. But in urban locations, greater densities make interaction with the natural environment harder to achieve. As people crowd together in cities, buildings are built more closely together, reducing access to daylight and fresh air, but increasing problems of noise and pollution, security and privacy. Active systems, especially artificial lighting and air-conditioning, make it possible to overcome these problems by controlling the internal environment independently of external conditions. While this reduces the need for the external fabric of the building to act as an environmental filter, it comes at a cost of increased energy consumption and contributes to greenhouse gas emission when that energy is generated from fossil fuels.

Throughout the 20th century, the rapid growth of well-serviced cities gave a perception that there would always be endless quantities of energy and water that could be used to create clean and healthy buildings. But today, the natural limits of the environment to provide energy and water and to absorb emissions are being recognized. If renewable sources can be developed, harvesting the available sunlight, wind or geothermal energy, then consumption can continue without causing damage to the natural environment. However, a more likely scenario is that fossil fuel use will continue for many years to come, or that electricity will be generated using nuclear power, thereby replacing one form of pollution with another. But even if renewable energy sources are not developed, it is possible to design buildings that are less reliant on external supply and are instead able to harvest ambient energy from the surrounding environment.

Ecological architecture

Creating buildings in ways that respect the natural environment is sometimes known as *ecologically sustainable design* (ESD) or *ecological architecture*.[1] The idea that architecture can be ecological suggests that levels of comfort can be provided for occupants through interaction with the surrounding ecosystem. It also suggests that buildings can be made to protect and conserve the natural ecosystems in which they are located. While the science of ecology has much to offer in terms of understanding the flows of energy, air and water in the natural environment, it is somewhat ironic to apply the term to architecture. The word *ecology*, used to describe the relationship between an organism and its environment, was derived from the Greek word *oikos*, meaning house. Ecology is the study of the way natural environments, such as oceans or forests, are able to support life.

In this sense, architecture is always ecological, creating a habitable environment for the occupants of built space. But in ecological terms, many of the practices currently used for making and operating buildings are not sustainable in the long term. What the science of ecology shows is that ecosystems depend upon a regular input of energy from the sun, which enables materials to be recirculated as they are transformed at various points throughout the system by chemical reactions. To make architecture that is ecological is to recognize the need for ongoing energy use, but to source that energy directly from the surrounding environment or from renewable sources. It is also to recognize the need for using materials that come from renewable or sustainable sources and that can either be reused or recycled in buildings, or disposed of in ways that do not harm the natural environment.[2]

As we have seen, many of the levels of energy, air and water already available in the natural environment are at levels that are fairly close to ideal for human life. Modification of the external environment is needed either to protect from extremes of cold, wind or rain, or to provide control over internal conditions independent of the time of day or the season. One of the principal roles of the architect is to arrange the form and fabric of buildings to make the best use of the natural environment, resolving, for example, the need for sunlight, daylight and ventilation in relation to issues such as privacy or view. Since building fabric alone cannot provide precise environmental control, architects have traditionally resorted to experience or precedent, rules of thumb, or codes and regulations. Where precise control is needed, consultants such as lighting designers and HVAC engineers are usually employed, often resulting in energy-intensive, air-conditioned buildings. Today, an important new area of expertise is available in the form of environmental engineers, many of whom have built upon knowledge and experience with building services, who are able to integrate passive and active systems to improve energy efficiency and help make buildings more sustainable.

Making better use of ambient energy sources can improve environmental performance, but will also have a significant effect on the functional and experiential aspects of built space. The way buildings work depends on providing suitable environmental conditions for the people inside and, although artificial systems can supplement or modify passive systems, they can never fully replace the need for human interaction with the natural environment. Buildings will often work better if the people within feel a sense of connection to the world outside. Similarly, while dramatic effects can be achieved through the use of artificial systems, the aesthetic impact of the way buildings capture the ever-changing patterns of energy, air and water will always form an essential part of architecture. For architecture to be ecological, it must help to create a human ecology, a living environment that respects and responds to the sensory interaction between the human body and the natural environment.

The aim of this book has been to explore ways in which various aspects of the natural environment – sun, heat, light, sound, air and water – are used to create conditions within the built environment, using either passive or active systems. Given that the construction and maintenance of buildings contributes significantly to the environmental impact of cities, there is a great potential for improvement – and much to be done. Architects must work with planners, engineers, builders and clients to create buildings that recognize the basic human need to interact with, as well as to protect, the natural environment.

DESIGN HEAT LOSS/GAIN RATE

A design heat loss/gain rate provides a useful figure to compare the rate at which a particular design, taking into account both size of building and materials used for construction, will either lose or gain heat to the surrounding environment. This provides a measure of the amount of energy needed to run appliances such as heaters or air-conditioners that are used to maintain a temperature difference between inside and out. The calculations are based on a hypothetical dwelling in London, with floor, wall and window areas and construction types as shown in the table overleaf. U values can be found in the *New Metric Handbook*, or in Szokolay, *Introduction to Architectural Science*, p. 254.

More detailed figures for heating degree days (HDD) and cooling degree days (CDD) can be obtained from the UK Met Office. These figures are based on maintaining particular temperature levels inside a building while the outside temperature changes, taking into account both the duration and the amount of the temperature difference. For example, heating degree days with a base of 18°C assumes that a heater is used to maintain an interior temperature at that level, measuring both how long the heater is used for and how far the outside temperature falls below 18°C.

Element	Area (m²)	Construction	U value (W/m²°C)	A × U (W/°C)
Roof	100	Metal-deck roof, R1.5 bulk insulation, reflective foil laminate, gypsum plasterboard (upward heat flow)	0.55	55
Walls	200	Brick-veneer wall, reflective foil laminate, gypsum plasterboard	0.68	136
Windows	90	Timber frame, single 6 mm glass	5.0	450
Floor	100 (ground floor only)	Concrete slab on ground	0.62	62
Ventilation		0.36 × volume × air changes per hour (0.36 derived from specific heat capacity of air = 1300J/m³°C, ÷ 3600 to convert to Wh/m³°C; our example assumes 1 air change per hour and building height of 6m for a two-storey dwelling, i.e. volume 600m³)		216
Overall design heat loss/gain rate		Σ (A × U) + Ventilation heat loss/gain		919
Annual heating load		Heat loss/gain rate × Heating degree days × 0.0864 to convert to MJ e.g. Heating degree days for London = 4068, giving 323,000MJ Annual heating load can then be used to estimate annual energy costs by taking into account cost per unit of energy, which depends on fuel type (e.g. gas) and appliance efficiency. Similar calculations can also be performed for annual cooling load.		

LUMEN METHOD

The amount of light reaching work surfaces in a room depends not only on the brightness of the lamps and on the number of luminaires installed, but also on the way the luminaires and the room work together to direct light down onto work surfaces. For a selected luminaire, data should include a utilization factor table (such as the example overleaf) that gives the percentage of light reaching work surfaces depending on room geometry and surface colours.

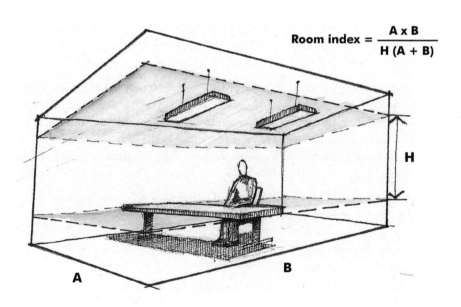

$$\text{Room index} = \frac{A \times B}{H (A + B)}$$

Typical utilization factor table

Ceiling	70	70	70	50	50	50	30	30	30
Wall	50	30	10	50	30	10	50	30	10
Floor	20	20	20	20	20	20	20	20	20
Room index	Utilization factor (%)								
0.75	18	16	14	18	15	14	17	15	13
1.00	22	19	17	21	19	17	20	18	17
1.25	24	22	20	23	21	19	23	21	19
1.50	26	24	22	25	23	21	24	22	21
2.00	29	26	24	28	26	24	27	25	23
2.50	30	28	26	29	27	26	28	27	25
3.00	32	30	28	30	29	27	29	28	27
4.00	33	32	30	32	31	29	31	29	28
5.00	34	33	32	33	32	31	31	31	30

For example, for a room that is 12×6 metres with luminaires installed 2 metres above the work surface, the room index would be $12 \times 6/2(12 + 6) = 2$; if surfaces were white ceiling (70 per cent), light-coloured walls (50 per cent) and dark carpet (20 per cent), the room index from the above table would be 29 per cent.

The amount of light reaching work surfaces also depends on the maintenance factor, which takes account of how frequently lamps and other surfaces are cleaned, and faulty lamps replaced, in relation to the amount of dust or other deposits likely to occur in the space. Maintenance factors can be found in the SLL Code for Lighting. For an office space with regular cleaning schedules, the maintenance factor will be around 95 per cent.

Calculating the number of luminaires required involves the following factors:
E = Average illuminance required on a work surface
A = Area of space
N = Number of luminaires installed
n = Number of lamps per luminaire
F = Luminous flux emitted by each lamp (from manufacturer's data)
Θ = Installed flux = $N \times n \times F$ (total amount of light from all lamps)
MF = Maintenance factor
UF = Utilization factor (see manufacturer's data for selected luminaire)

Amount of light on a surface =

(Amount of light from all lamps ÷ Area of space) × Loss factors

$E = (\Theta/A) \times MF \times UF$

$\quad = (N \times n \times F/A) \times MF \times UF$

Rearranging:

$N = E \times A / (F \times n \times UF \times MF)$

So, for the space described above, with dimensions 12×6 metres or 72 square metres, assuming:

- Required lighting level of 240 lux (recommended levels can be found in the SLL Code for Lighting)
- Output of 1800 lumens per lamp (actual output must be obtained from manufacturer's data)
- 2 lamps per luminaire (actual number depends on design of the luminaire)

N would be $240 \times 72 /(1800 \times 2 \times 0.95 \times 0.29) = 17.4$, rounding up to 18 lamps. These should be arranged evenly throughout the space, either in a regular (grid) or irregular pattern. The distance between lamps should not exceed the space-to-mounting height ratio (SHR) recommended by the manufacturer. This is normally around 1.5; that is, spacing should be no more than 1.5 times the height of the luminaire above the work plane, or $1.5 \times 2 = 3$ metres in the example given.

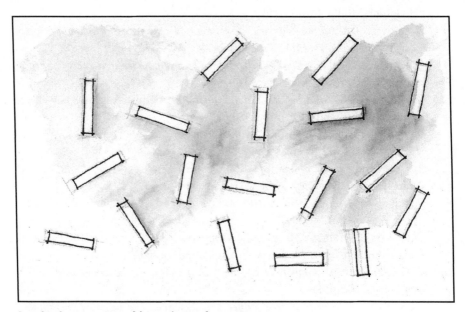

Luminaires arranged in an irregular pattern

NOTES

Introduction: Technology and environment

1   JM Fitch, *American Building: The forces that shape it*, Houghton Mifflin, Boston, 1948. The book was first published with two parts, looking at historical and environmental forces; these later became separated into two volumes, with the second (environmental) volume recently republished: JM Fitch with W Bobenhausen, *American Building: The environmental forces that shape it*, Oxford University Press, New York, 1999.

2   Fitch, *American Building, Volume 2: The environmental forces that shape it*, Houghton Mifflin, Boston, 1972, p. 1.

3   J Rykwert, *On Adam's House in Paradise: The idea of the primitive hut in architectural history*, MIT Press, Cambridge, MA, 1981.

4   L Fernández-Galiano, *Fire and Memory: On architecture and energy*, trans. G Cariño, MIT Press, Cambridge, MA, 2000.

5   JA Tainter, *The Collapse of Complex Societies*, Cambridge University Press, Cambridge, UK and New York, 1988.

6   J Diamond, *Guns, Germs and Steel: A short history of everybody for the last 13,000 years*, Vintage, London, 1998.

7   J Greenland, *Foundations of Architectural Science: Heat, light, sound*, University of Technology, Sydney, 1991, p. 8/14.

8   A Rapoport, *House Form and Culture*, Prentice-Hall, Englewood Cliffs, NJ, 1969.

9   D Hawkes & W Forster, *Architecture, Engineering and Environment*, Laurence King Publishers in association with Arup, London, 2002.

10 P Galison & E Thompson (eds), *The Architecture of Science*, MIT Press, Cambridge, MA, 1999.

11 SK Szokolay, *Introduction to Architectural Science: The basis of sustainable design*, Architectural Press, Oxford, 2003.

12 K Daniels, *Low-tech, Light-tech, High-tech: Building in the information age*, trans. E Schwaiger, Birkhäuser, Basel and Boston, 1998.

13 See W Bijker, T Hughes & T Pinch (eds), *The Social Construction of Technological Systems: New directions in the sociology and history of technology*, MIT Press, Cambridge, MA, 1987; also W Bijker & J Law (eds), *Shaping Technology/ Building Society: Studies in sociotechnical change*, MIT Press, Cambridge, MA, 1992.

14 TP Hughes, *Networks of Power: Electrification in Western society, 1880– 1930*, The Johns Hopkins University Press, Baltimore, 1983; D MacKenzie & J Wajcman (eds), *The Social Shaping of Technology: How the refrigerator got its hum*, Open University Press, Milton Keynes and Philadelphia, 1985.

15 E Schatzberg, 'Culture and technology in the city: Opposition to mechanized street transportation in late-nineteenth-century America', in MT Allen & G Hecht (eds), *Technologies of Power: Essays in honor of Thomas Parke Hughes and Agatha Chipley Hughes*, MIT, Cambridge, MA and London, 2001, pp. 57–94.

16 S Hoy, *Chasing Dirt: The American pursuit of cleanliness*, Oxford University Press, Oxford and New York, 1995.

17 L Mumford, *Technics and Civilization*, Harcourt Brace, New York, 1934.

18 R Mark, *Light, Wind, and Structure: The mystery of the master builders*, MIT Press, Cambridge, MA, 1990.

19 R King, *Brunelleschi's Dome: The story of the great cathedral in Florence*, Pimlico, London, 2005.

20 CD Elliott, *Technics and Architecture: The development of materials and systems for buildings*, MIT Press, Cambridge, MA, 1992.

21 R Banham, *The Architecture of the Well-tempered Environment*, Architectural Press, London, 1969.

22 D Hawkes, *The Environmental Imagination*, Routledge, London and New York, 2006; D Hawkes & W Forster, *Architecture, Engineering and Environment*, Laurence King, London, 2002; D Hawkes, J McDonald & K Steemers, *The Selective Environment*, Spon Press, New York, 2001; D Hawkes, *The Environmental Tradition: Studies in the architecture of environment*, Spon, London, 1996.

23 G Baird, *The Architectural Expression of Environmental Control Systems*, Spon Press, New York, 2001.

24 S Giedion, *Mechanization Takes Command: A contribution to anonymous history*, Oxford University Press, New York, 1948. See also JE Crowley, *The Invention of Comfort: Sensibilities & design in early modern Britain & early America*, Johns Hopkins University Press, Baltimore, 2001.

25 TF Tierney, *The Value of Convenience: A genealogy of technical culture*, State University of New York Press, Albany, 1993.

26 B Latour, 'Where are the missing masses? The sociology of a few mundane artifacts', in W Bijker & J Law (eds), *Shaping Technology/Building Society:*

Studies in sociotechnical change, MIT Press, Cambridge, MA, 1992, pp. 225–58. See also B Latour, 'Technology is society made durable', in J Law (ed.), *A Sociology of Monsters: Essays on power, technology, and domination*, Routledge, London and New York, 1991, pp. 103–31.

27   E Scarry, *The Body in Pain: The making and unmaking of the world*, Oxford University Press, New York and Oxford, 1985.

28   Scarry, *The Body in Pain*, pp. 38–39.

29   E Shove, *Comfort, Cleanliness and Convenience: The social organization of normality*, Berg, Oxford, 2003, p. 2.

30   J Appleton, *The Experience of Landscape*, John Wiley, London, 1975.

31   K Steemers & MA Steane (eds), *Environmental Diversity in Architecture*, Spon Press, London and New York, 2004.

32   SE Rasmussen, *Experiencing Architecture*, MIT Press, Cambridge, MA, 1964.

33   Steemers & Steane (eds), *Environmental Diversity in Architecture*.

34   J Pallasmaa, *The Eyes of the Skin: Architecture and the senses*, Academy Editions, London, 1996.

35   RE Cytowic, *The Man who Tasted Shapes*, Abacus, London, 1994.

36   L Heschong, *Thermal Delight in Architecture*, MIT Press, Cambridge, MA, 1979, pp. 24–30.

Chapter 1: Sun

1   A Latour (ed.), *Louis I. Kahn: Writings, lectures, interviews*, Rizzoli, New York, 1991.

2   When facing south in the northern hemisphere, the sun appears to travel from left to right, which may explain the Western preference for writing in the same direction.

3   B Rudofsky, *Architecture without Architects: A short introduction to non-pedigreed architecture*, University of New Mexico Press, Albuquerque, 1987 (1964).

4   V Richardson, *New Vernacular Architecture*, Laurence King, London, 2001.

5   RO Philips, *Sunshine and Shade in Australasia*, CSIRO, Australia, 1992.

6   R Knowles, *Sun Rhythm Form*, MIT Press, Cambridge, MA, 1981.

7   It is possible to extend the sunshade so that it provides full shading not just for the summer solstice, but for a slightly longer period, or 'shading season'. Choosing the maximum angle for the (May–July) sun path, for example, will extend the full shading for a period of about 12 weeks; but it must be remembered that every day of increased shading after the solstice will mean an extra day before the solstice as well – because of the symmetry of the sun path, the solstice will always occur at the midpoint of the shading season.

8   T Galloway, *Solar House: A guide for the solar designer*, Architectural Press, Oxford, 2004.

Chapter 2: Heat

1   L Heschong, *Thermal Delight in Architecture*, MIT Press, Cambridge, MA, 1979.

2   J Pallasmaa, *The Eyes of the Skin: Architecture and the senses*, Academy Editions, London, 1996, p. 41.

3   Engineers and architects sometimes use the term *coolth* to describe a potential source of heat loss. Conceptually, coolth is negative heat; coolth emanating from an object to cool a body is in fact heat radiating from the body to the object.

4   T Shachtman, *Absolute Zero and the Conquest of Cold*, Houghton Mifflin, Boston, 1999. Instead of using a scale that starts at this zero point (the Kelvin scale), it is more convenient to use a scale based on recognized and useful phenomena – the freezing and boiling points of water. The Celsius scale sets the freezing point of water at 0°C, and the boiling point of water at 100°C, with each degree being 1/100 of the temperature difference between these two events.

5   SJ Pyne, *Vestal Fire: An environmental history, told through fire, of Europe and Europe's encounter with the world*, University of Washington Press, Seattle, 1997, p. 41.

6   Fernández-Galiano, *Fire and Memory*, p. 233.

7   CD Elliott, *Technics and Architecture: The development of materials and systems for buildings*, MIT Press, Cambridge, MA, 1992, p. 280.

8   Fernández-Galiano, *Fire and Memory*, p. 251.

9   Fernández-Galiano, *Fire and Memory*, p. 241.

10   On the significance of the second law of thermodynamics to energy use in buildings, see also P Graham, *Building Ecology: First principles for a sustainable built environment*, Blackwell Science, Oxford, 2002.

11   The specific heat capacity is defined as the amount of energy needed to raise the temperature of one kilogram of a given material by one degree Celsius.

12   K_{total} is also described as the overall air-to-air thermal conductance, to indicate that surface effects have been included.

13   Tuff, P & Adler, D (eds), *New Metric Handbook*, Architectural Press, London, 1979.

Chapter 3: Light

1   MS Millet, *Light Revealing Architecture*, Van Nostrand Reinhold, New York, 1996.

2   M Frascari, 'The *lume materiale* in the architecture of Venice', *Perspecta 24*, autumn 1988, pp. 136–45.

3   MA Steane, 'Environmental diversity and natural lighting strategies', in K Steemers & MA Steane (eds), *Environmental Diversity in Architecture*, Spon Press, London and New York, 2004, pp. 159–78.

4   M Wigginton, *Glass in Architecture*, Phaidon, London, 1996.

5   C Alexander, S Ishikawa & M Silverstein, with M Jacobson, I Fiksdahl-King & S Angel, *A Pattern Language: Towns, buildings, construction*, Oxford University Press, New York, 1977, p. 526.

6   J Speirs, M Major & A Tischhauser, *Made of Light: The art of light and architecture*, Birkhäuser, Basel, 2004, pp. 26–33.

7   Alexander et al., *A Pattern Language*, pp. 746–49.

8   R Banham, *The Architecture of the Well-tempered Environment*, Architectural Press, London, 1969.

9   I Bentley, A Alcock, P Murrain, S McGlynn & G Smith, *Responsive Environments: A manual for designers*, The Architectural Press, London, 1985.

10   Actual lighting levels from the sky will be mostly brighter than this – lighting levels from a clear sky and bright sun can be in the order of 100,000 lux – but the minimum level indicates reliability, avoiding the need to turn on artificial lights if daylight falls below this for an extended period of time.

11   RG Hopkinson, P Petherbridge & J Longmore, *Daylighting*, Heinemann, London, 1966.

12   G Bachelard, *The Psychoanalysis of Fire*, trans. ACM Ross, Beacon Press, Boston. 1964.

13   D Nye, *Consuming Power: A social history of American energies*, MIT Press, Cambridge, MA, 1997; D Nye, *Electrifying America: Social meanings of a new technology*, MIT Press, Cambridge, MA, 1990.

14   WE Bijker, TP Hughes & TJ Pinch (eds), *The Social Construction of Technological Systems: New directions in the sociology and history of technology*, MIT Press, Cambridge, MA, 1987.

15   Recommended illuminance levels for a range of tasks can be found in the Society of Light and Lighting (SLL) Code for Lighting.

16   Luminous flux can be thought of as the 'brightness' of a lamp, although brightness is ambiguous as it can also be used to describe the amount of light that reaches the eye from a particular source.

17   D Clements-Croome (ed.), *Creating the Productive Workplace*, E & FN Spon, London and New York, 2000.

Chapter 4: Sound

1   W Ong, *Orality and Literacy: The technologizing of the word*, Routledge, London and New York, 2002.

2   D MacKenzie & J Wajcman (eds), *The Social Shaping of Technology: How the refrigerator got its hum*, Open University Press, Milton Keynes and Philadelphia, 1985.

3   K Steemers & MA Steane (eds), *Environmental Diversity in Architecture*, Spon Press, London and New York, 2004, p. 12.

4   J Pallasmaa, *The Eyes of the Skin: Architecture and the senses*, Academy Editions, London, 1996, p. 35.

5   Steemers & Steane (eds), *Environmental Diversity in Architecture*, p. 12.
6   E Thompson, *The Soundscape of Modernity: Architectural acoustics and the culture of listening in America, 1900–1933*, MIT Press, Cambridge, MA, 2004.
7   T Lewers, 'The reverential acoustic', in Steemers & Steane (eds), *Environmental Diversity in Architecture*, pp. 143–56.

Chapter 5: Air

1   CD Elliott, *Technics and Architecture: The development of materials and systems for buildings*, MIT Press, Cambridge, MA, 1992, pp. 273–79.
2   J Rykwert, *The First Moderns: The architects of the eighteenth century*, MIT Press, Cambridge, MA, 1980.
3   B Colomina, 'The medical body in modern architecture', in C Davidson (ed.), *Anybody*, Anyone Corp., New York; MIT Press, Cambridge, MA, 1997, pp. 228–39.
4   Elliott, *Technics and Architecture*, pp. 310–17.
5   Elliott, *Technics and Architecture*, pp. 308–10.
6   SJ Pyne, *Vestal Fire: An environmental history, told through fire, of Europe and Europe's encounter with the world*, University of Washington Press, Seattle, 1997, p. 51.
7   Elliott, *Technics and Architecture*, pp. 318–19.
8   D Arnold, 'The evolution of modern office buildings and air-conditioning', *ASHRAE Journal*, 41/6, June 1999, pp. 40–54; D Arnold, 'Air-conditioning in office buildings after World War II: the first century of air-conditioning', *ASHRAE Journal*, 41/7, July 1999, pp. 33–41.
9   Usually, an exchange of heat involves a change of temperature, as a hotter object cools down when it comes into contact with a cooler object, which warms up. However, at phase changes, substances can absorb or release heat without changing temperature. More importantly, the amount of heat that can be absorbed or released is relatively large. A saucepan full of water, for example, can be brought to the boil after only a few minutes, but will take many times that for it to boil away completely, converting from water at 100°C to steam at the same temperature. This can be extremely beneficial, since temperature changes can affect the rate of heat loss or heat gain and thus the efficiency of any heat exchange system.
10   In service-intensive buildings such as hospitals or laboratories, this figure can be several times greater. For details on air-conditioning design, including design stage estimates, see R Parlour, *Building Services: A guide to integrated design: engineering for architects*, Integral Publishing, Sydney, 2000.
11   G Baird, *The Architectural Expression of Environmental Control Systems*, Spon Press, New York, 2001.
12   Shove, *Comfort, Cleanliness and Convenience.*

13   The airflow can also be used to minimize the possibility of condensation occurring from cooling the air within the space.

14   On geothermal cooling, see also Elliott, *Technics and Architecture*, p. 314.

Chapter 6: Water

1   J-P Goubert, *The Conquest of Water: The advent of health in the industrial age*, trans. A Wilson, Princeton University Press, Princeton, NJ, 1989.

2   R Schwartz Cowan, *More Work for Mother: The ironies of household technology from the open hearth to the microwave*, Basic Books, New York, 1983.

3   B Colomina, 'The medical body in modern architecture', in C Davidson (ed.), *Anybody*, Anyone Corp., New York; MIT Press, Cambridge, MA, 1997, pp. 228–39.

4   A Loos, 'Plumbers', trans. HF Mallgrave, in N Lahiji & DS Friedman (eds), *Plumbing: Sounding modern architecture*, Princeton Architectural Press, New York, 1997, pp. 15–19.

5   M Wigley, *White Walls, Designer Dresses: The fashioning of modern architecture*, MIT Press, Cambridge, MA, 1995.

6   Le Corbusier, *Towards a New Architecture*, trans. F Etchells, Architectural Press, London, 1946, p. 89.

7   Le Corbusier, *Towards a New Architecture*, pp. 114–15.

8   AT Friedman, 'Domestic differences: Edith Farnsworth, Mies van der Rohe, and the gendered body', in C Reed (ed.), *Not at Home: The suppression of domesticity in modern art and architecture*, Thames & Hudson, London, 1996, pp. 179–92.

9   On the design of facilities for bathing, see A Kira, *The Bathroom: Criteria for design*, Center for Housing and Environmental Studies, Cornell University, Ithaca, NY, 1966.

10   M Douglas, *Purity and Danger: An analysis of concepts of pollution and taboo*, Routledge & Kegan Paul, London, 1966.

11   Goubert, *The Conquest of Water*, p. 171.

12   See, for example, S Hoy, *Chasing Dirt: The American pursuit of cleanliness*, Oxford University Press, Oxford & New York, 1995.

13   A Corbin, *The Foul and the Fragrant: Odour and the social imagination*, trans. M Kochan, R Porter & C Prendergast, Macmillan, London, 1996 (1986), p. 97.

Chapter 7: Fire

1   See also D Bass, 'Towering inferno: the metaphoric life of building services', *AA Files* 30, Autumn 1995, pp. 26–34.

2   On the behaviour of crowds in fire, see also P Ball, *Critical Mass: How one thing leads to another*, William Heinemann, London, 2004, pp. 174–80.

3   D Von Drehle, *Triangle: The fire that changed America*, Atlantic Monthly Press, New York, 2003.

Chapter 8: Ecological design

1   J Steele, *Ecological Architecture: A critical history 1900 – today*, Thames & Hudson, London, 2005.

2   J Fernandez, *Material Architecture: Emergent materials for innovative buildings and ecological construction*, Architectural Press, Boston, 2006; C Kibert, J Sendzimir & G Guy, *Construction Ecology: Nature as the basis for green buildings*, Taylor & Francis, New York, 2001; B Berge, *The Ecology of Building Materials*, trans. F Henley with H Liddell, Butterworth-Heinemann, Oxford, 2001; B Lawson, *Building Materials, Energy and the Environment: Towards ecologically sustainable development*, Royal Australian Institute of Architects, Canberra, 1996.

FURTHER READING

Introduction: Technology and environment

Allen, E, *How Buildings Work: The natural order of architecture*, 3rd edn, Oxford University Press, Oxford and New York, 2005.

Barbara, A & Perliss, A, *Invisible Architecture: Experiencing places through the sense of smell*, Skira, Milan, 2006.

Basalla, G, *The Evolution of Technology*, Cambridge University Press, Cambridge and New York, 1988.

Bijker, W, Hughes, TP & Pinch, TJ (eds), *The Social Construction of Technological Systems: New directions in the sociology and history of technology*, MIT Press, Cambridge, MA, 1987.

Bijker, W & Law, J (eds), *Shaping Technology/Building Society: Studies in sociotechnical change*, MIT Press, Cambridge, MA, 1992.

Elliott, CD, *Technics and Architecture: The development of materials and systems for buildings*, MIT Press, Cambridge, MA, 1992.

Fitch, JM with Bobenhausen, W, *American Building: The forces that shape it*, Oxford University Press, New York, 1999.

Greenland, J, *Foundations of Architectural Science: Heat, light, sound*, University of Technology, Sydney, 1991.

Hawkes, D, *The Environmental Imagination*, Routledge, London and New York, 2006.

Hughes, T, *Human-built World: How to think about technology and culture*, University of Chicago Press, Chicago and London, 2004.

Hyde, R, *Climate Responsive Design: A study of buildings in moderate and hot humid climates*, E & FN Spon, New York, 2000.

Mitcham, C, *Thinking through Technology: The path between engineering and philosophy*, University of Chicago Press, Chicago, 1994.

Olgyay, V, *Design with Climate: Bioclimatic approach to architectural regionalism*, Princeton University Press, Princeton, NJ, 1963.

Pacey, A, *Meaning in Technology*, MIT Press, Cambridge, MA, 1999.

Pallasmaa, J, *The Eyes of the Skin: Architecture and the senses*, Academy Editions, London, 1996.

Rosenbrock, H, *Machines with a Purpose*, Oxford University Press, Oxford and New York, 1990.

Rothenberg, D, *Hand's End: Technology and the limits of nature*, University of California Press, Berkeley, 1993.

Shove, E, *Comfort, Cleanliness and Convenience: The social organization of normality*, Berg, Oxford, 2003.

Steemers K & Steane MA (eds), *Environmental Diversity in Architecture*, Spon Press, London and New York, 2004.

Szokolay, SK, *Introduction to Architectural Science: The basis of sustainable design*, Architectural Press, Oxford, 2003.

www.yourhome.gov.au

Chapter 1: Sun

British Standards Institution, *Basic Data for the Design of Buildings: Sunlight*, Draft for Development DD67, British Standards Institution, Milton Keynes, 1980.

Greenland, J & Szokolay, S, *Passive Solar Design in Australia*, RAIA Education Division, Canberra, 1985.

Hollo, N, *Warm House Cool House: Inspirational designs for low-energy housing*, Choice Books, Sydney, 1995.

Knowles, R, *Ritual House: Drawing on nature's rhythms for architecture and urban design*, Island Press, Washington, DC, 2006.

Phillips, RO, *Sunshine and Shade in Australasia*, CSIRO, Australia, 1992.

Schittich, C (ed.), *Solar Architecture: Strategies, visions, concepts*, Birkhäuser, Basel, 2003.

Chapter 2: Heat

Fernández-Galiano, L, *Fire and Memory: On architecture and energy*, trans. G Cariño, MIT Press, Cambridge, MA, 2000.

Heschong, L, *Thermal Delight in Architecture*, MIT Press, Cambridge, MA, 1979.

Segre, G, *A Matter of Degrees: What temperature reveals about the past and future of our species, planet, and universe*, Viking, New York, 2002.

Shachtman, T, *Absolute Zero and the Conquest of Cold*, Houghton Mifflin, Boston, 1999.

Szokolay, S, *Thermal Design of Buildings*, RAIA Education, Canberra, 1987.

Tutt, P & Adler, D (eds), *New Metric Handbook*, London: Architectural Press, 1979.

Chapter 3: Light

Meyers, V, *Designing with Light*, Laurence King, London, 2006.

Millet, MS, *Light Revealing Architecture*, Van Nostrand Reinhold, New York, 1996.

Speirs, J, Major, M & Tischhauser, A, *Made of Light: The art of light and architecture*, Birkhäuser, Basel, 2004.

Daylight

Ander, GD, *Daylighting Performance and Design*, John Wiley, Hoboken, NJ, 2003.

Baker, N, Fanchiotti, A & Steemers, K (eds), *Daylighting in Architecture: A European reference book*, James & James, London, 1993.

Baker, N & Steemers, K, *Daylight Design of Buildings*, James & James, London, 2002.

British Standards Institution, *BS 8206-02: Lighting for Buildings – Part 2: Code of practice for daylighting*, British Standards Institution, London, 1992.

Bell J and Burt W, *Designing Buildings for Daylight*, Building Research Establishment, Garston, 1995.

Hopkinson, RG, Petherbridge, P & Longmore, J, *Daylighting*, Heinemann, London, 1966.

Köster, H, *Dynamic Daylight Architecture: Basics, systems, projects*, Birkhäuser, Basel; Momenta, London, 2002.

Phillips, D, *Daylighting: Natural Light in Architecture*, Architectural Press, Oxford, 2004.

BRE Digest 309, 310: Estimating Daylight in Buildings, Building Research Establishment, Garston, 1986.

Artificial light

Department of Employment and Industrial Relations, Working Environment Branch, *Artificial Light at Work*, Australian Government Publishing Service, Canberra, 1984.

Egan, MD & Olgyay, VW, *Architectural Lighting*, McGraw-Hill, Boston, 2002.

Colour

Ball, P, *Bright Earth: The invention of colour*, Viking, London, 2001.

Linton, H, *Color in Architecture: Design methods for buildings, interiors, and urban spaces*, McGraw-Hill, New York and London, 2003.

Mack, G, *Colours: Rem Koolhaas/OMA, Norman Foster, Alessandro Mendini*, Birkhäuser, Boston, 2001.

McCown, J, *Colors: Architecture in detail*, ed. OR Ojeda, photography Paul Warchol, Rockport Publishers, Gloucester, MA, 2004.

Moor, A, *Colours of Architecture*, Mitchell Beazley, London, 2006.

Toy, M (ed.), *Colour in Architecture*, Architectural Design profile 120, Academy Editions, London, 1996.

Chapter 4: Sound

Brooks, CN, *Architectural Acoustics*, McFarland & Co., Jefferson, NC, 2003.

Egan, MD, *Architectural Acoustics*, McGraw-Hill, New York, 1988.

Grueneisen, P, *Soundspace: Architecture for sound and vision*, Birkhäuser, Basel; Springer, London, 2003.

Long, M, *Architectural Acoustics*, Academic, Oxford, 2005.

Mehta, M, Johnson, J & Rocafort, J, *Architectural Acoustics: Principles and design*, Prentice Hall, Upper Saddle River, NJ, 1999.

Parkin, PH & Humphreys, HR, *Acoustics, Noise and Buildings*, Faber, London, 1969.

Chapter 5: Air

Air-conditioning

Ackerman, M, *Cool Comfort: America's romance with air-conditioning*, Smithsonian Institute Press, Washington, DC, 2002.

Bachelard, G, *Air and Dreams: An essay on the imagination of movement*, Dallas Institute of Humanities and Culture, Dallas, 1988.

Baird, G, *The Architectural Expression of Environmental Control Systems*, Spon Press, New York, 2001.

Banham, R, *The Architecture of the Well-tempered Environment*, 2nd edn, University of Chicago Press, Chicago, 1984.

Cooper, G, *Air-conditioning America: Engineers and the controlled environment, 1900–1960*, Johns Hopkins University Press, Baltimore, 1998.

System design

Chadderton, DV, *Building Services Engineering*, Spon Press, New York, 2004.

Daniels, K, *The Technology of Ecological Building: Basic principles and measures, examples and ideas*, Birkhäuser Verlag, Basel and Boston, 1997.

Daniels, K, *Advanced Building Systems: A technical guide for architects and engineers*, trans. E Schwaiger, Birkhäuser Verlag, Basel, 2003.

Hawkes, D & Forster, W, *Architecture, Engineering and Environment*, Laurence King, London, 2002.

Heerwagen, D, *Passive and Active Environmental Controls: Informing the schematic designing of buildings*, McGraw-Hill, New York, 2004.

Lechner, N, *Heating, Cooling, Lighting: Design methods for architects*, John Wiley, New York, 2000.

McQuiston, FC, Parker, JD & Spitler Danvers, JD, *Heating, Ventilating, and Air Conditioning: Analysis and design*, John Wiley & Sons, Hoboken, NJ, 2005.

Parlour, RP, *Building Services: A guide to integrated design: engineering for architects*, Integral Publishing, Sydney, 2000.

Natural ventilation

Allard, F (ed.), *Natural Ventilation in Buildings: A design handbook*, James & James, London, 1998.

Awbi, HB, *Ventilation of Buildings*, Taylor & Francis, New York, 2003.

Battle McCarthy Consulting Engineers, *Wind Towers*, Academy Editions, Chichester, UK, 1999.

Clements-Croome, D (ed.), *Naturally Ventilated Buildings: Buildings for the senses, economy and society*, E & FN Spon, London and New York, 1997.

Chapter 6: Water

Bachelard, G, *Water and Dreams: An essay on the imagination of matter*, Pegasus Foundation, Dallas, 1983.

Ball, P, *Life's Matrix: A biography of water*, Farrar, Straus & Giroux, New York, 2000.

Corbin, A, *The Foul and the Fragrant: Odour and the social imagination*, trans. M Kochan, R Porter & C Prendergast, Macmillan, London, 1996 (1986).

Douglas, M, *Purity and Danger: An analysis of concepts of pollution and taboo*, Routledge & Kegan Paul, London, 1966.

Goubert, J-P, *The Conquest of Water: The advent of health in the industrial age*, trans. A Wilson, Princeton University Press, Princeton, NJ, 1989.

Hoy, S, *Chasing Dirt: The American pursuit of cleanliness*, Oxford University Press, Oxford and New York, 1995.

Illich, I, *H_2O and the Waters of Forgetfulness*, Boyars, London, 1986.

Kaika, M, *City of Flows: Modernity, nature, and the city*, Routledge, New York, 2005.

Kira, A, *The Bathroom: Criteria for design*, Center for Housing and Environmental Studies, Cornell University, Ithaca, NY, 1966.

Lahiji, N & Friedman, DS (eds), *Plumbing: Sounding modern architecture*, Princeton Architectural Press, New York, 1997.

Schwartz Cowan, R, *More Work for Mother: The ironies of household technology from the open hearth to the microwave*, Basic Books, New York, 1983.

Swyngedouw, E, *Social Power and the Urbanization of Water: Flows of power*, Oxford University Press, New York, 2004.

Wright, L, *Clean and Decent: The fascinating history of the bathroom & the water closet, and of sundry habits, fashions & accessories of the toilet, principally in Great Britain, France, & America*, Routledge & Kegan Paul, London, 1960.

Chapter 7: Fire

Bass, D, 'Towering inferno: the metaphoric life of building services', *AA Files* 30, Autumn 1995, pp. 26–34.

Egan, MD, *Concepts in Building Firesafety*, Wiley, New York, 1978.

Parlour, RP, *Building Services: A guide to integrated design: engineering for architects*, Integral Publishing, Sydney, 2000.

Patterson, J, *Simplified Design for Building Fire Safety*, Wiley, New York, 1993.

Ramsay, GC & Rudolph, L, *Landscape and Building Design for Bushfire Areas*, CSIRO Publishing, Melbourne, 2003.

Stollard, P & Abrahams, J, *Fire from First Principles: A design guide to building fire safety*, 2nd edn, E & FN Spon, London and New York, 1995.

Chapter 8: Ecological design

Baker, N & Steemers, K, *Energy and Environment in Architecture: A technical design guide*, E & FN Spon, New York, 2000.

Berge, B, *The Ecology of Building Materials*, trans. F Henley with H Liddell, Butterworth-Heinemann, Oxford, 2001.

Buchanan, P, *Ten Shades of Green: Architecture and the natural world*, Architectural League of New York, distributed by WW Norton, New York, 2005.

Edwards, B, *Towards a Sustainable Architecture: European directives and building design*, Butterworth Architecture, Oxford and Boston, 1999.

Edwards, B, *Green Architecture: An international comparison*, Academy Editions, London, 2001.

Farmer, J, *Green Shift: Towards a green sensibility in architecture*, Architectural Press, Boston, 1999.

Fernandez, J, *Material Architecture: Emergent materials for innovative buildings and ecological construction*, Architectural Press, Boston, 2006.

Graham, P, *Building Ecology: First principles for a sustainable built environment*, Blackwell Science, Oxford, 2002.

Hagan, S, *Taking Shape: A new contract between architecture and nature*, Architectural Press, Boston, 2001.

Kibert, C, Sendzimir, J & Guy, GB, *Construction Ecology: Nature as the basis for green buildings*, Taylor & Francis, New York, 2001.

Lawson, B, *Building Materials, Energy and the Environment: Towards ecologically sustainable development*, Royal Australian Institute of Architects, Canberra, 1996.

Melet, E, *Sustainable Architecture: Towards a diverse built environment*, NAI Publishers, Rotterdam, 1999.

Mobbs, M, *Sustainable House: Living for our future*, Choice Books, Sydney, 1998.

Porteous, C, *The New Eco-architecture: Alternatives from the Modern movement*, Spon Press, New York, 2001.

Sassi, P, *Strategies for Sustainable Architecture*, Taylor & Francis, New York, 2005.

Smith, P, *Architecture in a Climate of Change: A guide to sustainable design*, Butterworth-Heinemann, Oxford, 2001.

Steele, J, *Ecological Architecture: A critical history*, Thames & Hudson, London, 2005.

Thomas, R (ed.), *Environmental Design: An introduction for architects and engineers*, Taylor & Francis, London and New York, 2005.

Williamson, T, Radford, A & Bennetts, H, *Understanding Sustainable Architecture*, Spon Press, London, 2003.

Wines, J, *Green Architecture*, Taschen, Cologne and London, 2000.

Wines, J, *The Art of Architecture in the Age of Ecology*, Taschen, Cologne and London, 2000.

Woolley, T, Kimmins, S, Harrison, R & Harrison, P, *Green Building Handbook: A companion guide to building products and their impact on the environment*, Spon Press, London, 2000.

Zeiher, L, *The Ecology of Architecture: A complete guide to creating the environmentally conscious building*, Whitney Library of Design, New York, 1996.

INDEX